T0313963

THE volumes of the University of Michigan Studies are published by authority of the Executive Board of the Graduate School of the University of Michigan. A list of the volumes thus far published or arranged for is given at the end of this volume.

University of Michigan Studies

HUMANISTIC SERIES

VOLUME XI

CONTRIBUTIONS TO THE HISTORY OF SCIENCE

PART I. ROBERT OF CHESTER'S LATIN TRANSLATION OF THE ALGEBRA OF AL–KHOWARIZMI

THE MACMILLAN COMPANY
NEW YORK · BOSTON · CHICAGO · DALLAS
ATLANTA · SAN FRANCISCO

MACMILLAN & CO., LIMITED
LONDON · BOMBAY · CALCUTTA
MELBOURNE

THE MACMILLAN CO. OF CANADA, LTD.
TORONTO

PLATE I.

CODEX VINDOBONENSIS 4770, Fol. 1ᵃ.

ROBERT OF CHESTER'S

LATIN TRANSLATION

OF THE

ALGEBRA OF AL–KHOWARIZMI

WITH AN INTRODUCTION, CRITICAL NOTES
AND AN ENGLISH VERSION

BY

LOUIS CHARLES KARPINSKI

UNIVERSITY OF MICHIGAN

New York

THE MACMILLAN COMPANY

LONDON : MACMILLAN AND COMPANY LIMITED

1915

Printed and bound by CPI Group (UK) Ltd, Croydon, CR0 4YY

Norwood Press
J. S. Cushing Co. — Berwick & Smith Co.
Norwood, Mass., U.S.A.

Paperback ISBN: 978-0-472-75151-8

PREFACE

FROM the point of view of the history of science, no justification is needed for the publication of a mathematical text of the twelfth century, for the available material representing this period is meagre. A wider acquaintance with Robert of Chester's Latin translation of Al-Khowarizmi's Arabic treatise on algebra will perhaps contribute to a more just estimate of the services rendered to science by the Arabs. In the English version I have not attempted to give a literal translation of the Latin, but rather to express the thought in a phraseology which the modern student of mathematics will find easy of comprehension; by consulting the Latin text and footnotes the reader will be able to examine Robert of Chester's own words. For the convenience of readers interested in the text I have added a Latin Glossary in which are noted many variations from the usage of classical writers. In the Introduction I have presented a study of the significance of the treatise in the history of mathematics, and a description of the manuscripts upon which the text is based.

It is a pleasure to express indebtedness to Professor David Eugene Smith for having suggested the work; to Mr. George A. Plimpton for the generous use of his unique mathematical library; to the librarians of the libraries of Vienna and Dresden for photographic reproductions of manuscripts containing this text; to the librarian of Columbia University for the loan of the Scheybl manuscript, and to the librarian of the Cleveland Public Library for the use of works from the John G. White collection. I am much indebted also to colleagues of the University of Michigan, particularly in the Department of Latin and the Department of Mathematics.

I am under special obligation to Mr. William H. Murphy for making possible this publication.

LOUIS C. KARPINSKI.

ANN ARBOR, MICHIGAN,
November 1, 1915.

v

CONTENTS

PLATES

INTRODUCTION

CHAPTER I

ALGEBRAIC ANALYSIS BEFORE AL-KHOWARIZMI

ARABIC contributions to science have, in the past, been somewhat neglected by historians. More recent studies are recognizing our indebtedness to Mohammedan scholars, who kept the embers of learning aglow while Europe was in the darkness of the Middle Ages. Much of our knowledge of Greek mathematics comes to us from Arabic sources; the early Latin versions were frequently based upon Arabic texts rather than the Greek originals. Similarly, Hindu arithmetic and astronomy were transmitted to Europe by Islam. The services of the Arabs to science were not limited to the preservation and transmission of the learning of other nations. They made independent contributions in many fields.

Among these achievements is the Arabic algebra of Al-Khowarizmi, which for centuries enjoyed wide popularity in the original, and for further centuries extended its popularity through translations and adaptations. A study of the content of this work is an excursion into mediæval thought. By a study of the text in a form as nearly like the original as possible, we discover the reason for its long-continued appeal to the Occidental as well as the Oriental mind, its interest for the Englishman, the German, and the Italian, as well as for the Arab. Even to-day teachers of elementary mathematics may find this book fruitful in suggestion: the geometric solutions of quadratic equations presented by the Arabic writer more than a thousand years ago may be used with profit in our classrooms.

Simple equations of the first degree in one unknown, of the type $ax = b$, are found in the oldest mathematical text-book which we possess, the Ahmes papyrus of about 1700 B.C., which was published with a German translation by Eisenlohr.[1] This Egyp-

---

[1] A. Eisenlohr, *Ein mathematisches Handbuch der alten Aegypter*, Leipzig, 1877; *Facsimile of the Rhind Mathematical Papyrus in the British Museum*, with preface by E. A. Wallis Budge, London, 1898.

tian document presents not only first degree equations together with symbols for the unknown quantity and for the operations of addition and subtraction, but also shows traces of a study of simultaneous linear equations some two thousand years before the Christian Era. Later, but still before the golden age of Greek mathematics, the quadratic equation appears in Egypt. The problems found involve simultaneous quadratic equations, thus: [1]

"Another example of the distribution of a given area into squares. If you are told to distribute 100 square ells (units of area) over two squares so that the side of one shall be $\frac{3}{4}$ of the side of the other: please give me each of the unknowns." The solution follows by assuming the side of one square to be unity, and the other $\frac{3}{4}$. The sum of these areas is $\frac{25}{16}$, of which the root is $\frac{5}{4}$. The root of 100 is 10; 10, then, is to the required side as $\frac{5}{4}$ is to 1, whence one side is 8 and the other 6. The algebraical equivalent of this geometrical problem is, evidently,

$$x^2 + y^2 = 100,$$
$$y = \tfrac{3}{4}\, x.$$

Noteworthy also is the fact that a symbol for square root occurs in the discussion of these problems.

The solution above leads to the number relation, $6^2 + 8^2 = 10^2$, which connects directly with the simpler form, $3^2 + 4^2 = 5^2$, and to the same relation other problems of this kind reduce.[2] This makes connection, of course, with the so-called Pythagorean theorem that the sum of the squares on the sides of a right triangle equals the square on the hypotenuse. Even though the Egyptians had no logical proof for this proposition, their familiarity with it is well established. In the time of Plato, and for some centuries afterwards, the Egyptians were famed as surveyors, and the principle stated seems to have been applied by them in laying out right angles by means of a long rope knotted at equal intervals. Two pegs situated three units apart are set out along the line to which it is desired to draw a perpendicular. From one peg

[1] M. Cantor, *Vorlesungen über Geschichte der Mathematik*, Vol. I, third edition (Leipzig, 1907), pp. 95–96. To this work we shall refer as Cantor, I (3), and to the other volumes similarly.

See also Max Simon, *Geschichte der Mathematik im Altertum* (Berlin, 1909), pp. 41–42; Schack-Schackenburg, *Zeitschrift für Aegyptische Sprache*, Vol. XXXVIII (1900), pp. 135–140, and Vol. XL, pp. 65–66.

[2] Cantor, I (3), p. 96.

an arc is swung with a radius of four units, while from the other end an arc is swung with a radius of five units. The intersection is connected with the peg from which the shorter arc is swung, forming thus a right angle with the desired line, for in any triangle with sides in the ratio three to four to five, a right angle lies opposite the longest side.

The Pythagorean theorem was applied also in India, before the time of Pythagoras, in the construction of altars. With this theorem as developed in the Apastamba Sulba Sutras,[1] the rules for altar construction, are associated careful approximations of square root, pure quadratic equations, and even, as Milhaud has shown,[2] the possible solution of the complete quadratic equation,

$$ax^2 + bx = c.$$

The ancient Babylonians, furthermore, constructed tables of squares and cubes. Such tables are found upon the famous tablets of Senkereh,[3] which are contemporary with the Ahmes papyrus. Application of these quadratic numbers to problems similar to those of Egypt already mentioned has not been discovered, but the fact is evident that such tables were a step toward the study of quadratic equations. Cantor[4] shows that the ancient Hebrews were probably familiar with the 3, 4, 5 right triangle. In China, too, students mathematically inclined had come upon this number relation,[5] and evidently were studying quadratic numbers.

Familiarity of Greek mathematicians with the geometrical solution of quadratic equations in the time of Pythagoras is now well established.[6] Hippocrates (fifth century B.C.) writing on the quadrature of the lunes, in an attempt to square the circle, assumes a construction which is equivalent to the solution of the equation,[7]

$$x^2 + \sqrt{\frac{3}{2}}\, ax = a^2.$$

[1] Bürk, *Das Āpastamba-Śulba-Sutra*, *Zeitschrift der deutschen Morgenländischen Gesellschaft*, Vol. LV (1901), pp. 543–591, and Vol. LVI (1902), pp. 327–391.

[2] G. Milhaud, *La Géométrie d'Apastamba*, *Revue générale des Sciences*, Vol. XXI (1910), pp. 512–520; see also T. L. Heath, *The thirteen books of Euclid's Elements* (3 vols., Cambridge, 1908), Vol. I, pp. 352–364. We shall refer to this latter work as Heath's *Euclid*.

[3] Cantor, I (3), pp. 25–31.   [4] Cantor, I (3), p. 49.

[5] Cantor, I (3), pp. 181 and 679–680.   [6] Heath's *Euclid*, Vol. I, pp. 386–387, 403.

[7] T. L. Heath, *Diophantus of Alexandria, A study in the history of Greek algebra*, second edition (Cambridge, 1910), p. 63; more detailed in Rudio, *Der Bericht des Simplicius über die Quadraturen des Antiphon und des Hippokrates* (1907), p. 58, and same author in the *Bibliotheca Mathematica*, Vol. III, third series (1902), pp. 7–42.

Several propositions of Euclid present geometrical equivalents of the solution of various types of quadratic equations, not involving negative coefficients, and further study of similar problems appears in Euclid's *Data*. Of this nature are the fifth, sixth, and eleventh propositions of the second book of the *Elements* and the twenty-seventh, twenty-eighth, and twenty-ninth of the sixth book, and problems 84, 85, 86, and others of the *Data*.[1] Problem 84, for example, reads:

"If two straight lines include a given area in a given angle, and the excess of the greater over the less is given, then each of them is given."

This corresponds to the equations:

$$xy = k^2$$
$$x - y = a.$$

The two following problems (85 and 86) correspond to the simultaneous quadratic equations:

$$xy = k^2,$$
$$x + y = a,$$

and

$$xy = k^2,$$
$$x^2 - y^2 = a^2.$$

The eleventh proposition of the second book of the *Elements* furnishes the solution of the equation

$$x^2 + ax = a^2$$

or even more general,

$$x^2 + ax = b^2.$$

As this so well illustrates the geometrical solution, it is given in full, following Heath's *Euclid*.

Book II of the *Elements of Euclid*, Proposition 11

"To cut a given straight line so that the rectangle contained by the whole and one of the segments is equal to the square on the remaining segment.

"Let *AB* be the given straight line; thus it is required to cut *AB* so that the rectangle contained by the whole and one of the segments is equal to the square on the remaining segment.

---

[1] References and citations from the *Elements* are to Heath's *Euclid* and the *Data* (Greek and Latin) edited by H. Menge, Leipzig, 1896, being Vol. VI of *Euclidis opera omnia*, ed. Heiberg et Menge. An English translation of the *Data* is found in the numerous editions of *The Elements of Euclid* by Simson. The numbering of the problems is slightly different in the two versions.

"For let the square $ABDC$ be described on AB (I. 46); let AC be bisected at the point E, and let BE be joined ; let CA be drawn through to F, and let EF be made equal to BE ; let the square FH be described on AF, and let GH be drawn through to K.

"I say that AB has been cut at H so as to make the rectangle contained by AB, BH equal to the square on AH.

"For, since the straight line AC has been bisected at E, and FA is added to it, the rectangle contained by CF, FA together with the square on AE is equal to the square on EF. (II. 6.)

"But EF is equal to EB ; therefore the rectangle CF, FA together with the square on AE is equal to the square on EB.

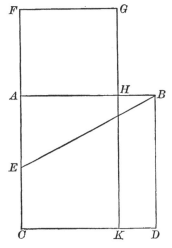

"But the squares on BA, AE are equal to the square on EB, for the angle at A is right (I. 47) ; therefore the rectangle CF, FA together with the square on AE is equal to the squares on BA, AE.

"Let the square on AE be subtracted from each ; therefore the rectangle CF, FA which remains is equal to the square on AB.

"Now the rectangle CF, FA is FK, for AF is equal to FG ; and the square on AB is AD ; therefore FK is equal to AD.

"Let AK be subtracted from each ; therefore FH which remains is equal to HD.

"And HD is the rectangle AB, BH, for AB is equal to BD ; and FH is the square on AH ; therefore the rectangle contained by AB, BH is equal to the square on HA.

"Therefore the given straight line AB has been cut at H so as to make the rectangle contained by AB, BH equal to the square on HA. Q. E. F."

The ordinary algebraical solution of the corresponding equation

$$x^2 + ax = a^2, \qquad \text{from } a(a-x) = x^2,$$

parallels this geometrical demonstration.

To complete the square in the left-hand member, $\dfrac{a^2}{4}$ is added to both members. This corresponds to marking the point E on the figure, for then the square on BE equals $a^2 + \dfrac{a^2}{4}$ or $AB^2 + AE^2$.

Extracting the square root of both members, we have, algebraically,

$$x + \frac{a}{2} = + \sqrt{\frac{5\,a^2}{4}},$$

the negative sign being disregarded. The right-hand member corresponds to the line BE and the left-hand member to EF, which is equal to BE.

Algebraically we proceed by subtracting $\frac{a}{2}$ from both members,

giving
$$x = \sqrt{\frac{5\,a^2}{4}} - \frac{a}{2}.$$

This corresponds to the line AF in the figure which is $BE - AE$.

Analytical solution of the quadratic equation appears quite definitely in the works of Heron of Alexandria, who flourished about the beginning of the Christian Era. Heron states in effect that given the sum of two line segments and their product then each of the segments is known.[1] However, he goes farther than any work of Euclid in applying this to a numerical example,

$$144\, x(14 - x) = 6720.$$

Without putting this into the form of an equation, Heron states that the approximate value of x is $8\frac{1}{2}$, and this evidently indicates an analytical solution. The geometrical garb is absolutely discarded in a problem in the *Geometry* doubtfully attributed to Heron.[2] The problem is to compute the diameter of a circle given the sum of the area, the circumference, and the diameter, summing an area and lengths, entirely contrary to Greek usage. The form of the result, practically

$$x = \frac{\sqrt{(154 \times 212 + 841)} - 29}{11},$$

indicates that the equation

$$\tfrac{11}{14}\,x^2 + \tfrac{29}{7}\,x = 212$$

was put in the form

$$121\,x^2 + 638\,x = (212)(154).$$

Somewhat similar problems[3] in which lines and areas are summed appear in Greece in the period between Heron and Diophantus (about 250 A.D.) as well as in the works of the latter. One of these problems is to find a square whose area and perimeter together equal 896 ($x^2 + 4\,x = 896$). The solution pro-

---

[1] Heron, *Metrica*, ed. Schöne (Leipzig, 1903), pp. 148–151.

[2] Cantor, I (3), p. 405; Heron, *Geometria*, ed. Hultsch (Berlin, 1864), p. 133; *Heronis Opera*, ed. Heiberg, Vol. IV, *Geometria*, p. 381; Heath, *Diophantus*, pp. 63–64.

[3] Heiberg and Zeuthen, *Ueber einige Aufgaben der unbestimmten Analytik, Bibliotheca Mathematica*, Vol. VIII, third series, pp. 118–134. The date is conjectural. See Heath, *Diophantus*, pp. 118–121.

ceeds in the ordinary manner by adding to 896 the square of half the coefficient of x and then taking the square root of this sum. From this is subtracted one-half the coefficient of x, giving the side 28. Four other problems of this series deal with right triangles having rational sides and hypotenuse, in which the sum of the area and perimeter is to equal a given number. If a, b are the sides, c the hypotenuse, S the area, r radius of the inscribed circle, and $s = \frac{1}{2}(a + b + c)$, then the solution depends upon the following formulæ:

$$S = rs = \tfrac{1}{2} ab, \ r + s = a + b, \ c = s - r.$$
$$\left.\begin{matrix} a \\ b \end{matrix}\right\} = \frac{r + s \pm \sqrt{(r + s)^2 - 8\, rs}}{2}.$$

In the sixth book of the *Arithmetica* Diophantus treats rational right triangles in which the area plus or minus one side is a given number, or the area plus or minus the sum of two sides or one side and the hypotenuse, is a given number. Again, such a problem appears in an algebraic work by Shojā ben Aslam, Abū Kāmil, an Arabic writer of the tenth century.[1]

In respect to analysis Diophantus is the greatest name among the Greeks. Recently it has been established that he flourished in the third century of the Christian Era, when Greek supremacy in mathematics was waning. No doubt whatever exists that this Alexandrian was familiar with the analytical solution of the various forms of quadratic equations, neglecting negative roots and, of course, imaginary roots, which did not receive serious treatment for more than a millennium after Diophantus. The three types of complete quadratic equations, involving only positive coefficients, are the following:

$$ax^2 + bx = c,$$
$$ax^2 + c = bx,$$
$$ax^2 = bx + c.$$

All three types appear in the *Arithmetica* of Diophantus, not systematically treated but solved as incidental to the solution of other problems. In fact, after dealing with the solution of equations of the form

$$ax^m = bx^n,$$

[1] Suter, *Die Abhandlung des Abū Kāmil Shoğā' b. Aslam "über das Fünfeck und Zehneck,"* *Bibliotheca Mathematica*, Vol. X, third series (1910–11), pp. 15–42.

Diophantus makes the following explicit statement regarding his intention of writing a systematic treatise on the quadratic equation:[1]

> "This should be the object aimed at in framing the hypotheses of propositions, that is to say, to reduce the equations, if possible, until one term is left equal to one term; but I will show you later how, in the case also where two terms are left equal to one term, such a problem is solved."

So far as we know this promise was never fulfilled.

An equation of the first type is presented by the sixth problem of the sixth book, and this we reproduce from Heath:[2]

> "6. To find a right-angled triangle such that the area added to one of the perpendiculars makes a given number.
>
> "Given number 7, triangle $(3x, 4x, 5x)$.
>
> "Therefore $6x^2 + 3x = 7$."
>
> "In order that this might be solved it would be necessary that (half coefficient of $x)^2 +$ product of coefficient of x^2 and absolute term should be a square: but $(1\frac{1}{2})^2 + 6 \cdot 7$ is not a square. Hence we must find, to replace $(3, 4, 5)$, a right-angled triangle such that
>
> "$(\frac{1}{2}$ one perpendicular$)^2 + 7$ times area $=$ a square;"

and the subsequent work leads to the equation, $84x^2 + 7x = 7$, $x = \frac{1}{4}$; and the solution $(6, \frac{7}{4}, \frac{25}{4})$.

In the following problem (VI. 7) the value of x is given as $\frac{1}{3}$ for the equation

$$84x^2 - 7x = 7,$$

which equation is of the third type when the negative term is transposed after the manner of Diophantus. The equations

$$630x^2 + 73x = 6, \qquad x = \frac{1}{18},$$
$$630x^2 - 73x = 6, \qquad x = \frac{6}{35},$$
$$630x^2 + 81x = 4, \qquad x = \frac{4}{105},$$

and
$$630x^2 - 81x = 4, \qquad x = \frac{1}{6},$$

occur in the next four problems (VI. 8–11). Another problem of the third kind (IV. 39) is of especial interest because the rule is given for solving this type:[3]

> "To find three numbers such that the difference of the greatest and the middle has to the difference of the middle and the least a given ratio, and also the sum of any two is a square."

[1] Heath, *Diophantus*, p. 131.

[2] Heath, *Diophantus*, pp. 228–229.

[3] Heath, *Diophantus*, pp. 197–198.

The discussion leads to the inequality, $2\,m^2 > 6\,m + 18$, which, since only integral solutions are desired, explains the use of 7 as the approximate square root of 45, in the following paragraph:

" When we solve such an equation, we multiply half the coefficient of x into itself, — this gives 9, — then multiply the coefficient of x^2 into the units, — $2 \cdot 18 = 36$, — add this last number to the 9, making 45, and take the side [square root] of 45, which is not less than 7 ; add half the coefficient of x, — making a number not less than 10, — and divide the result by the coefficient of x^2 ; the result is not less than 5."

Of the second type, $ax^2 + c = bx$, Diophantus gives several illustrations, requiring frequently only the approximate value of the root. Problems of this kind are the following:

$$72\,m > 17\,m^2 + 17, \qquad m \text{ not greater than } \tfrac{67}{17}, \qquad \text{(V. 10)}$$
$$19\,m^2 + 19 > 72\,m, \qquad m \text{ not less than } \tfrac{66}{19}, \qquad \text{(V. 10)}$$
$$m^2 + 60 > 22\,m, \qquad m \text{ not less than } 19, \qquad \text{(V. 30)}$$
$$m^2 + 60 < 24\,m, \qquad m \text{ not greater than } 21, \qquad \text{(V. 30)}$$

and
$$172\,x = 336\,x^2 + 24, \qquad \text{(VI. 22)}$$

of which the statement is made that the root is not rational, and in the same problem

$$78848\,x^2 - 8432\,x + 225 = 0,$$

which has the rational root $\tfrac{25}{448}$.

Commentaries on the *Arithmetica* began to appear very early. Probably the most interesting commentary from the modern point of view was the one written in the late fourth or early fifth century by Hypatia, the daughter of Theon of Alexandria. Unfortunately her writings are all lost, although there is ground for the belief [1] that some remarks made by Michael Psellus (eleventh century) concerning Egyptian arithmetic and algebra were based on her commentary. She came naturally by her mathematical ability ; her father Theon wrote a commentary on Ptolemy's Almagest and makes in this the earliest known reference to Diophantus.

Cossali, [2] writing in 1797 on the history of algebra, conjectures that the step from the geometrical to the analytical form of solution took place in the period between Euclid and Diophantus.

---

[1] *Diophantus*, ed. P. Tannery (Leipzig, 1893), Vol. II, pp. 37–38 ; Heath, *Diophantus*, pp. 18, 41.

[2] *Origine, trasporto in Italia, primi progressi in essa dell' algebra* (Parma, 1797), Vol. I, pp. 87–91.

Now the Arabic Book of Chronicles[1] (987 A.D.) states that the astronomer Hipparchus (second century B.C.) wrote a treatise on algebra, and Cantor[2] inclines to the belief that there actually was such a work. No trace, however, has been found of it, and the probability is that Hipparchus did not write any systematic treatise on algebra or on quadratic equations. The word "algebra" indeed is Arabic in its origin and the use of it as a title goes back only to the time of our author, Mohammed ibn Musa. Nevertheless it is possible that Greek mathematicians of the time of Hipparchus did occupy themselves with problems of the kind in question, because this was a natural development out of the consideration of rational right triangles, as given by Pythagoras and Plato, in connection with the geometrical treatment of quadratic equations as given by Euclid. Quadratic equations connect even more directly with the application of areas, of Pythagorean origin, which is extensively treated by Euclid.

Some two centuries after the period of Diophantus, Āryabhaṭa, one of the earliest Hindu mathematicians of prominence, was born (476 A.D.). In the work of Āryabhaṭa as presented by Rodet[3] we find the solution of a quadratic equation assumed in the rule for finding the number of terms of an arithmetic series when the sum, difference, and first term are given. Nor does Āryabhata in India stand alone in the study of analysis, as Diophantus does in Greece. Brahmagupta of Ujjain, the centre of Indian learning, wrote on algebra in the early part of the seventh century and gave a rule[4] for the solution of quadratic equations:

"To the absolute number multiplied by the [coefficient of the] square, add the square of half the [coefficient of the] unknown, the square root of the sum, less half the [coefficient of the] unknown, being divided by the [coefficient of the] square, is the unknown."

In formula this corresponds to the solution

$$x = \frac{\sqrt{(b/2)^2 + ac} - b/2}{a}$$

[1] *Das Mathematiker-Verzeichniss im Fihrist des Ibn Abi Ja'kub an-Nadim*, translated by H. Suter, *Abhandl. z. Geschichte der Mathematik*, Vol. 6 (Leipzig, 1892), p. 22, and note, pp. 54–55. Suter holds that there is some error in the text, and this seems probable.

[2] *Geschichte*, Vol. I (3), pp. 362–363.

[3] *Leçons de Calcul d'Āryabhata, Journal Asiatique*, seventh series, Vol. XIII (1879), pp. 393–434.

[4] Colebrooke, *Algebra, with Arithmetic and Mensuration, from the Sanskrit of Brahmegupta and Bháscara* (London, 1817), p. 347; Cantor, I (3), p. 625.

of the equation,

$$ax^2 + bx = c.$$

Contemporary with Al-Khowarizmi is the Hindu writer Mahavi-racarya, whose arithmetical and algebraical work has been translated into English by M. Rangacarya.[1] Rules are given in the sections devoted to algebra for the three types of complete quadratic equations. A peculiarity of the treatment is that the unknown quantity and the square root of the unknown appear, rather than the unknown and its square. The significance of the work is that it shows a persistence of interest in algebra in India from the time of Āryabhaṭa. Three centuries later Bhaskara (b. 1114 A.D.), another Hindu mathematician, made important contributions to the advance of the science.

The brief survey which we have given of the study of algebra before the time of Mohammed ibn Musa does not at all purpose to present the sources from which the great Arab drew his inspiration. Greece undoubtedly took mathematical ideas from Egypt, as Rodet[2] some years ago pointed out with reference to algebra. Even more definite evidence is presented by the Greek use of unit fractions as well as by the references to Egyptian mathematics which were made by Plato and Herodotus, and much later by Michael Psellus. Babylon and Greece were constantly exchanging ideas;[3] a striking proof of this is the Greek use of sexagesimal fractions. India, too, was not out of touch with these neighbors to her west. Especially in the fields of religion and the closely associated astrology we have abundant evidence not only of interchange of ideas between the East and the West but also of the recurrence in mediæval times of ideas advanced by more ancient civilizations. Yet we need to notice that we are dealing with the independent appearances of algebraic ideas, and that the mathematics of Egypt, Babylon, China, Greece, and India was developing from within. Algebra is not, as often assumed, an

[1] M. Rangācārya, *The Ganita-Sāra-Sangraha of Mahāvīrācārya* (Madras, government press, 1912). Professor D. E. Smith gave a brief preliminary report of the work in the *Bibliotheca Mathematica*, Vol. IX, third series, pp. 106–110.

[2] L. Rodet, *Sur les notations numériques et algébriques antérieurement au XVIᵉ siècle* (Paris, 1881), pp. 43–51.

[3] F. Cumont, *Babylon und die griechische Astronomie, Neue Jahrbücher f. das klassische Altertum . . .*, Vol. 27 (1911), pp. 1–10; *The Oriental Religions in Roman Paganism* (Chicago, 1911); and *Astrology and Religion among the Greeks and Romans* (New York, 1912).

artificial effort of human ingenuity, but rather the natural expression of man's interest in the numerical side of the universe of thought.   Tables of square and cubic numbers in Babylon; geometric progressions, involving the idea of powers, together with linear and quadratic equations in Egypt; the so-called Pythagorean theorem in India, and possibly in China, before the time of Pythagoras; and the geometrical solution of quadratic equations even before Euclid in Greece, are not isolated facts of the history of mathematics.   While they do indeed mark stages in the development of pure mathematics, this is only a small part of their significance.   More vital is the implication that the algebraical side of mathematics has an intrinsic interest for the human mind not conditioned upon time or place, but dependent simply upon the development of the reasoning faculty.   We may say that the study of powers of numbers, and the related study of quadratic equations, were an evolution out of a natural interest in numbers; the facts which we have presented are traces of the process of this evolution.

CHAPTER II

Al-Khowarizmi and his Treatise on Algebra

THE activity of the great Arabic mathematician Abu 'Abdallah Mohammed ibn Musa al-Khowarizmi marks the beginning of that period of mathematical history in which analysis assumed a place on a level with geometry; and his algebra gave a definite form to the ideas which we have been setting forth. The arithmetic of Al-Khowarizmi made known to the Arabs and, through an early twelfth-century translation, to Europeans also, the Hindu art of reckoning. Any consideration of the difficulties attending arithmetical operations with the Greek letter numerals,[1] or even with the Roman numerals, shows how essential the adoption of numerals with place value was for the development of the analytical side of mathematics; the way was being prepared also for the appearance of decimal fractions, many centuries later, and for logarithms, both indispensable tools of modern science. Quite as important as the arithmetic for the development of mathematics was the systematic treatise on algebra[2] which Mohammed ibn Musa gave to the world. This is the work of which we present the Latin translation made by Robert of Chester while living in Segovia in 1183 of the Spanish Era (1145 A.D.).[3]

Our chief source of information in regard to the life and the writings of our Arabic author is the *Book of Chronicles*[4] (*Kitab*

---

[1] Nine letters represent the units from 1 to 9, nine further letters the tens from 10 to 90, and nine others the hundreds. The thousands up to 900,000 are represented by the same letters with a kind of accent mark.

[2] The Arabic text as given in the MS. Hunt, 214, of the Bodleian library, a unique copy, was published with English translation by Frederic Rosen, *The Algebra of Mohammed ben Musa* (London, 1831), being one of the volumes published for the Oriental Translation Fund; a French translation of the chapter on Mensuration was published by Marre, based on Rosen's Arabic text, *Nouvelles Annales de Mathématiques*, Vol. 5 (1846), pp. 557–581, and also later revised in *Annali di Matemat.*, Vol. VII (1866), pp. 268–280.

[3] A preliminary note by the author concerning this translation appeared in the *Bibliotheca Mathematica*, third series, Vol. XI (1911), pp. 125–131, with the title, *Robert of Chester's Translation of the Algebra of Al-Khowarizmi*.

[4] The *Kitab al-Fihrist* was edited with notes by G. Flügel and published after Professor Flügel's death by J. Roediger and A. Mueller (Leipzig, 1871–1872). I quote from the German translation by H. Suter, *Das Mathematiker-Verzeichniss im Fihrist, Abhandlungen zur Geschichte der Mathematik*, Vol. VI, Leipzig, 1892.

al-Fihrist) by Ibn Abi Ya'qub al-Nadim. This work, which was completed about 987 A.D., gives biographies of learned men of all nations, together with lists of such of their works as were known to Al-Nadim. We quote the passage relating to our author, regarding whose life and activity we have only meagre information.

" AL-KHOWARIZMI "

"Mohammed ibn Musa, born in Khowarizm (modern Khiva), worked in the library of the caliphs under Al-Mamun. During his lifetime, and afterward, where observations were made, people were accustomed to rely upon his tables, which were known by the name *Sind-Hind* (Hindu *Siddhanta*). He wrote : *The book of astronomical tables* in two editions, the first and the second ; *On the sun-dial; On the use of the astrolabe ; On the construction of the astrolabe ; The Book of Chronology.*"

Neither the date of the birth nor the date of the death of Al-Khowarizmi has been definitely established. However, the fact, mentioned by Al-Nadim in the *Fihrist*, that he worked in the library of the caliph Al-Mamun, who reigned from 813 to 833 A.D., indicates the period of his literary activity. The introduction to the algebra, which is not found in the extant Latin translations, is here given in English (p. 45), following Rosen. This brings further evidence of the acquaintance with Al-Mamun, for Al-Khowarizmi states that the interest of the caliph in science encouraged him to write the treatise. The probability is that early in the reign of Al-Mamun our author began work upon the Hindu astronomical tables which, as the Fihrist account implies, brought him almost immediate fame. This stimulated him to undertake the work upon algebra, and the success of the second work induced him to write the treatise on arithmetic in which reference is made to the algebra. The height of his literary activity may reasonably be placed about 825 A.D.

The bibliography of Al-Nadim does not include four works from the hand of Al-Khowarizmi which have come to us. These are his arithmetic, his algebra, a work on the quadrivium, and an adaptation of Ptolemy's geography. To Sened ibn 'Ali, the Jew, whose biography immediately follows that of Mohammed ibn Musa, are ascribed works entitled, *On Increasing and Decreasing* (algebraical), *The Book of Algebra*, and *On the Hindu Art of Reckoning*. The probability is, as Suter points out,[1] that an inter-

---

[1] *Fihrist, loc. cit.*, pp. 62–63.

change has taken place here, although this must have been relatively early since Ibn al-Qifti,[1] who died in 1248 A.D., in his *Chronicles of the Learned*, gives the same account of Al-Khowarizmi as Al-Nadim. Furthermore, the author of the Fihrist knew of the algebra, for he mentions no less than three men as commentators on the algebra of Mohammed ibn Musa; Sinan ibn al-Fath of Harran, 'Abdallah ibn al-Hasan al-Saidanani, and Abu'l-Wefa al-Buzjani are credited with such commentaries.[2]

The arithmetic of Al-Khowarizmi has come down to us only in a Latin translation, and this survives in a unique copy belonging to the library of the University of Cambridge, published in 1857 by Prince Baldassare Boncompagni.[3] Several references[4] in this work to the writer's other book on arithmetic make it evident that the Al-Khowarizmi in question is the author of the algebra. Our word *algorism*, as well as the obsolete form *augrim*,[5] used by

---

[1] Suter, *loc. cit.*, pp. 62–63; Casiri, *Bibliotheca Arabico-Hispana Escurialensis* (Madrid, 1760–1770), I, pp. 427–428. Ibn al-Qifti mentions the arithmetic.

[2] Suter, *loc. cit.*, pp. 37, 36, and 39 respectively.

[3] *Trattati d' aritmetica* (Rome, 1857), I. *Algoritmi de numero indorum*, pp. 1–23, which is the arithmetic in question, and bears internal evidence that it is a translation from the Arabic; II. *Joannis Hispalensis liber algorismi de pratica arismetrice*, evidently an adaptation and enlargement of the preceding. Some other manuscripts of the second treatise are anonymous, while others ascribe it to Gerard of Cremona.

[4] One passage is in *Trattati*, I, pp. 2–3; Etiam patefeci in libro algebre et almucabalah, idest restaurationis et oppositionis, quod uniuersus numerus sit compositus, et quod uniuersus numerus componatur super unum. Vnum ergo inuenitur in uniuerso numero; et hoc est quod in alio libro arithmetice dicitur. Quia unum est radix uniuersi numeri, et est extra numerum. Radix numeri est, quare per eum inuenitur omnis numerus. . . . Reliquus autem numerus sine uno inueniri non potest. . . . Reliquus autem numerus indiget . . . uno: quare non potes dicere duo uel tria, nisi precedat unum. Nichil aliud est ergo numerus, nisi unitatum collectio. . . .

Inueni, inquit algorizmi, omne quod potest dici ex numero, et esse quicquid excedit unum usque in IX., id est quod est inter. IX. et unum, id est duplicatur unum et fiunt duo; et triplicatur idem unum, fiuntque tria, et sic in ceteris usque in IX. De inde ponuntur, X. in loco unius, et duplicantur, X. ac triplicantur, quemadmodum factum est de uno, fiuntque ex eorum duplicatione, XX, et triplicatione, XXX, et ita usque ad XC. Post hec redeunt C in loco unius, et duplicantur ibi atque triplicantur, quemadmodum factum est de uno et X; efficiunturque ex eis CC, et CCC, et cetera usque in DCCCC. Rursum ponuntur mille in loco unius; et duplicando et triplicando, ut diximus, fiunt ex eis II mila, et III et cetera usque in infinitum numerum, secundum hunc modum.

Compare with this our text, p. 66, lines 10–21.

Another passage is *Trattati*, I, p. 10. Etiam patefeci in libro, quod necesse est omni numero qui multiplicatur in aliquo quolibet numero, ut duplicetur unus ex eis secundum unitates alterius.

Compare with this our text, p. 90, lines 6–8.

[5] See my paper on *Augrim-stones*, *Modern Language Notes*, Vol. XXVII (1912), pp. 206–209; compare also the use of the term *Algaurizin* in our text, p. 102.

Chaucer, is derived from the use of the name Al-Khowarizmi in the opening sentence of the arithmetic, which reads, *Dixit algoritmi* or 'Algorithm says'; the word 'algebra' is derived from its use as a title by Al-Khowarizmi in the work which we are presenting.[1]   Up to the eighteenth century the common name for the new arithmetic with the ten figures of India, 1 2 3 4 5 6 7 8 9 0, was algorism, or in Latin *algorismus*.   Interesting also is the fact that a Spanish transliteration of Khowarizm, *guarismo*, is used for 'numerals,' corresponding to our ordinary use of 'ciphers.' Aside from Al-Khowarizmi's arithmetic in Latin translation, which was never widely used, the works which served to introduce the numerals into Europe were the *Carmen de Algorismo*,[2] in verse, written by Alexander de Villa Dei (about 1220), and the *Algorismus vulgaris*[3] by John of Halifax, commonly known as Sacrobosco (about 1250).   Both of these works were somewhat dependent upon Mohammed ibn Musa's arithmetic, and both continued in wide use for centuries.   Many manuscript copies of the *Carmen* are found in European libraries, and rather more of the *Algorismus vulgaris*. Even after the invention of printing Sacrobosco's Algorism was widely used for university instruction in arithmetic, many editions appearing in the fifteenth and sixteenth centuries.[4]   Detailed and extended commentary upon the work was given by Petrus de Dacia in 1291, in his lectures, evidently, and probably in a similar manner by many other university lecturers.

Another arithmetical treatise ascribed to Al-Khowarizmi is found in the *Liber ysagogarum Alchorismi in artem astronomi-cam a magistro A. compositus*.[5]   The principles of arithmetic, geometry, music, and astronomy are explained in five books, or chapters, and in two manuscripts there follow three books on

---

[1] See my note on *Algebra, Modern Language Notes*, Vol. XXVIII (1913), p. 93; compare also the use of the term in our text, p. 2.

[2] Published by J. O. Halliwell, *Rara Mathematica* (London, 1839).

[3] There were many early editions; see Curtze, *Petri Philomeni de Dacia in Algorismum vulgarem Johannis de Sacrobosco commentarius, una cum Algorismo ipso* (Ed. M. Curtze, Copenhagen, 1897.)

[4] See Smith, *Rara Arithmetica* (Boston, 1908), pp. 31–33; see also my article, *Jordanus Nemorarius and John of Halifax, American Mathematical Monthly*, Vol. XVII (1910), pp. 108–113.

[5] Nagl, in *Zeitschrift f. Math. u. Physik, Hist.-litt. Abth.*, XXXIV, pp. 129–146, 161–170; the first three books complete, and summary of the other two, by Curtze, *Ueber eine Algorismus-schrift des XII. Jahrhunderts* in *Abhandl. z. Gesch. d. math. Wissen.*, VIII, pp. 1–27; Haskins, in *The English Historical Review*, XXVI, p. 494.

astronomy. Three of the books which deal particularly with arithmetic have been published,[1] but no study has yet been made of the three books on astronomy. The writer A. is supposed to be Adelard of Bath, who was active at the time that this work was written; he translated the tables of Al-Khowarizmi. So far as these five books of the introduction are concerned, the work may well be a summary of the elementary teachings of Al-Khowarizmi according to the unknown writer's conception of them.

Al-Mas'udi (885–956 A.D.) in his *Meadows of Gold*[2] mentions Mohammed ibn Musa among the historians and chroniclers, basing his reference doubtless on the *Book of Chronology* above mentioned. Al-Biruni (973–1048 A.D.), whose work on India[3] has recently (1910) appeared in a second English edition, refers to the tables and the astronomical work of our author. No less than three works by Al-Biruni[4] are explanatory of works written by the distinguished mathematician and astronomer who was his fellow-countryman, both being from Khowarizm (or Khiva). Not only in the fields of astronomy, chronology, and mathematics did Mohammed ibn Musa achieve fame, but also as a geographer. His contribution in this field has been set forth by Nallino, who states, by way of conclusion to his article on *Al-Khowarizmi and his reconstruction of the geography of Ptolemy*,[5] that this geography is not a servile imitation of the Greek model, but an elaboration of Ptolemaic material made with more independence and ability than is displayed by any European writer of that period. The trigonometric tables, in Latin translation by Adelard of Bath, appear to be the earliest of Al-Khowarizmi's works to be known in Europe, and indeed one of the earliest mathematical treatises taken from the Arabic; it was translated in 1126 A.D. A study

---

[1] Curtze, *loc. cit.*; Haskins, *loc. cit.*, gives the incipit and explicit of the astronomical work: Quoniam cuiusque actionis quantitatem temporis spacium metitur, celestium motuum doctrinam querentibus eius primum ratio occurrit investiganda. . . . Divide quoque arcum diei per 12 et quod fuerit erunt partes horarum eius, si deus inveniri consenserit.

[2] *Les prairies d'or*, translated by Barbier de Meynard and Pavet de Courteille, I–IX (Paris, 1861–1877), I, 11.

[3] *Alberuni's India*, translation by C. E. Sachau (London, 1888, and second edition, London, 1910).

[4] Suter, *Der Verfasser des Buches Gründe der Tafeln des Chowārezmī, Bibliotheca Mathematica*, third series, Vol. IV, 1903, pp. 127–129.

[5] *Al-Ḫuwarizmi e il suo rifacimento della Geografia di Tolomeo, Atti della R. Accademia dei Lincei*, fifth series, *Memorie, Classe di scienze morali, storiche e filologiche*, Vol. II (1896), pp. 11–53.

of these tables, with extracts from the Latin text, was made by A. A. Björnbo[1] and the completed work, edited by Suter, has recently been published. Suter has found[2] that this translation is not of the original work by Al-Khowarizmi, but is based on a revision of that work made by Maslama al-Majriti (about 1000 A.D.). The work is of considerable importance for the history of trigonometry, for even though the tangent function which appears therein may be the addition of Maslama, yet the introduction of the sine function is certainly carried back as far as Al-Khowarizmi.

The combination of mathematical, geographical, and astronomical interests exhibited by Al-Khowarizmi renders plausible the hypothesis advanced by Suter,[3] on chronological grounds, that this Mohammed ibn Musa took part in the measurement of a degree of the earth's circumference which was made at the request of the caliph Al-Mamun. Some early Arabic chroniclers join the three sons of Moses, the Beni Musa, in this task. The oldest of these brothers was Mohammed and so he was called after the Arabic custom, Mohammed ibn Musa, meaning Mohammed, the son of Moses. Abu Ja'far is the prefix to his name, and this has been incorrectly given by numerous modern dictionaries[4] as the name of our author.

Concerning Mohammed ibn Musa's fame among the Arabs as an algebraist abundant evidence exists. Not only the commentaries cited bear witness to this fame but also the recurrent appearance for centuries of the numerical examples, $x^2 + 10\,x = 39$, $x^2 + 21 = 10\,x$, $3\,x + 4 = x^2$ and many others, which Al-Khowarizmi used. Some of the later authors, like Abu Kamil Shoja' ibn Aslam[5] (about 925 A.D.), explicitly acknowledge their indebtedness

---

[1] Björnbo, *Al-Chwarizmi's trigonometriske Tavler*, in *Festskrift til H. G. Zeuthen*, Copenhagen (1909), pp. 1–26; the sudden death of this brilliant student of mediæval mathematics is a great loss, for his systematic studies in this field were beginning to bear fruit in numerous and important publications.

[2] Suter, *Mémoires de l'Acad. Royale des Sciences et des Lettres de Danemark*, Series 7 (Lettres), Vol. 3, pt. 1 (Copenhagen, 1914). The dependence upon Maslama's version was noted by C. H. Haskins, *Adelard of Bath, The English Historical Review*, Vol. XXVI (1911), pp. 491–498.

[3] Suter, *Die Mathematiker und Astronomen der Araber und ihre Werke*, in *Abhandl. z. Gesch. d. math. Wissenschaften*, Vol. X (Leipzig, 1900), p. 20.

[4] *Century, Webster's, The New English Dictionary, Encyclopædia Britannica*.

[5] L. C. Karpinski, *The algebra of Abu Kamil Shoja'ben Aslam, Bibliotheca Mathematica*, Vol. XII, third series (1912), pp. 40–55; *The Algebra of Abu Kamil, American Mathematical Monthly*, Vol. XXI (1914), pp. 37–48.

to our author. Others, like the poet Omar ibn Ibrahim al-Khayyami (about 1045–1123 A.D.), familiar to English readers as Omar Khayyam, and Mohammed ibn al-Ḥasan al-Karkhi (died about 1029), do not consider it necessary to state the source of these problems and the proofs, which had become classic. Thus the equation

$$x^2 + 10x = 39$$

runs like a thread of gold through the algebras for several centuries, appearing in the algebras of the three writers mentioned, Abu Kamil, Al-Karkhi [1] and ʿOmar al-Khayyami,[2] and frequently in the works of Christian writers, centuries later, as we shall have occasion to note below.

Recognition of the fame of Al-Khowarizmi is to be found in the explicit statement by Ibn Khaldun (1332–1406) in his encylopædic work [3]: "The first who wrote upon this branch (algebra) was Abu ʿAbdallah al-Khowarizmi, after whom came Abu Kamil Shojaʿ ibn Aslam." Haji Khalfa, some two hundred years later, makes a similar statement. The *Chronicles of the Learned* by Ibn al-Qifti (d. 1282 A.D.) speak very highly of his ability as an arithmetician,[4] and a contemporary·of Al-Qifti, Zakariya ibn Moh. ibn Mahmud al-Qazwini, refers to him as translating the art of algebra for the Mohammedans.[5]

I have recently made a study [6] of the algebra of Abu Kamil, whose name we have seen associated by Arabic historians with that of Al-Khowarizmi. I have shown that he drew extensively upon the work of his predecessor, and further that Leonard of Pisa (1202 A.D.) drew, in turn, even more extensively from Abu Kamil. Thus the influence of the Khowarizmian was carried over to Italy as well as to the Arabic commentators and the European translators

---

[1] F. Woepcke, *Extrait du Fakhri* (Paris, 1853); A. Hochheim, *Die Arithmetik des Abu Bekr Muhammed ben Alhusein Alkarkhi* (Programm, Magdeburg, 1878), and *Kāfī fil Hisāb* (*Genügendes über Arithmetik*) (Halle, 1878–1880). The *Kāfī fil Hisāb* is an earlier work by Al-Karkhi which includes a treatment of algebra.

[2] F. Woepcke, *L'algèbre d'ʾOmar Alkhayyāmī* (Paris, 1851).

[3] MacGuckin de Slane, *Prolegomènes historiques d'Ibn Khaldoun, Notices et Extraits des Manuscrits de la Bibliothèque Imperiale et autres Bibliothèques* (Text, Vol. XIX and XX, Translation, Vol. XXI, 1868), Vol. XXI, p. 136.

[4] Casiri, *Bibliotheca Arabico-Hispana Escurialensis*, p. 427.

[5] Casiri, *loc. cit.*, p. 427; Cossali, *Origine, trasporto in Italia, primi progressi in essa dell' algebra* (Parma, 1797), Vol. I, p. 178.

[6] *The Algebra of Abu Kamil, loc. cit.*

of Abu Kamil's algebra. Two of these commentaries on Abu Kamil's work appeared in the tenth century and are listed in the Fihrist. The Persian Al-Istakhri,[1] known as The Reckoner, was the author of one, and 'Ali ibn Ahmed al-'Imrani [2] wrote the other. Neither commentary has come down to us. The author of the Latin translation upon which my study is based is not known, but the probability is that the translation was made about the time of Gerard of Cremona. Another commentary of uncertain date in Arabic by a Mohammedan Spaniard, Al-Khoreshi, is probably of later origin. In the fifteenth century a Hebrew translation was prepared by Mordechai Finzi of Mantua (about 1475); this is of especial interest since it bears internal evidence, in the terminology employed, that it was based upon a Spanish original. The author of the Spanish treatise, doubtless a Christian, is unknown. Ibn Khaldun mentions the commentary written by Al-Khoreshi, but no trace has yet been found of either Spanish translation or commentary. A unique copy of the Latin translation of the algebra is in Paris (Mss. Lat. 7377 A) and copies of the Hebrew translation are found in Paris and Munich.[3]

The widespread interest among Arabic scientists in the study of algebra is attested by the number of works upon the subject. Even as early as the tenth century besides Al-Khowarizmi there were three other writers[4] of sufficient prominence to warrant their appearance in the Fihrist. Abu 'Otman Sahl ibn Bishr. ibn Habib ibn Hani, a Jew, wrote an algebra which Al-Nadim states was praised by the Romans, and Aḥmed ibn Dā'ūd, Abū Ḥanifa, al-Dīnawarī (d. 895 A.D.) and Abu Yusuf al-Missisi also wrote treatises on the subject. These works are preserved only in title. In the thirteenth century Ibn al-Banna,[5] including in his arithmetic a brief exposition of algebra, employs the title, "algebra and almucabala," as given by Al-Khowarizmi, and follows the same peculiar order of the six types of quadratic equations. Al-Banna adds no numerical illustrations but states

---

[1] Suter, *Die Mathematiker und Astronomen der Araber*, p. 51.

[2] Suter, *loc. cit.*, pp. 56–57.

[3] Steinschneider, *Die Hebraeischen Uebersetzungen des Mittelalters* (Berlin, 1893), pp. 584–588; *Sitzungsber. d. Akad. d. Wissen.* (Vienna), *Phil.-hist. Klasse,* Vol. 149 (1905), p. 31.

[4] Suter, *Die Mathematiker*, pp. 15, 31, and 66 respectively.

[5] Suter, *Die Mathematiker*, pp. 162–163 ; A. Marre, *Le Talkhys d' Ibn Albanna, Atti dell' accad. pontif. dei nuovi Lincei,* Vol. XVII (1864), pp. 289–319.

simple general rules for the solution of equations. A separate work on the same subject by Ibn al-Banna is included among the Arabic manuscripts in Cairo. In modern times Arabs have used lithograph copies of an arithmetic and algebra by Mohammed ibn al-Hosein, Behā al-dīn al-ʿĀmilī (Behā ed-dīn), who died in 1622. Two editions [1] with a translation into the Persian language and commentary were published in the nineteenth century. Beha ed-din continues the use of some of Al-Khowarizmi's problems as well as the six types of quadratics.

Greek influence on Arabic geometry is revealed by the order of the letters employed on the geometrical figures. These letters follow the natural Greek order and not the Arabic order. The same is true of the order of the letters in the geometrical figures used by Al-Khowarizmi for verification of his solutions of quadratic equations. However, the somewhat ingenious hypothesis put forth by Cantor [2] that this fact shows that these demonstrations are from Greek sources is hardly tenable. The Arabs were much more familiar with and grounded in Euclid than are mathematicians to-day, and it was entirely natural in constructing new figures that they should follow the order of lettering to which they had grown accustomed in their study of Euclid. Until further and more definite historical evidence to the contrary is brought to light we must regard Al-Khowarizmi as the first to bring out sharply the parallelism between the analytical and geometrical solutions of quadratic equations.

The Arabic students of algebra included poets, philosophers and perhaps even kings. Omar Khayyam, whom we have mentioned among the writers on the subject, was too excellent a mathematician and too true a poet to woo the muse of algebra in verse. But in the library of the Escurial at Madrid there is preserved a poem treating of algebra, written by a native of Granada, Mohammed al-Qasim. Needless to say, the content, adapted to the exigencies of verse, does not compare with the

---

[1] Calcutta, 1812, and Constantinople, 1851–1852 ; Arabic with German translation by Nesselmann, Berlin, 1843 ; French translation, A. Marre, Paris, 1864.

[2] Cantor, I (3), pp. 724–725 ; Simon has noted the weakness of Cantor's argument in the article, *Zu Hwarizmi's hisāb al ǧabr wal muqābala*, *Archiv der Mathematik und Physik*, third series, XVIII (1911), pp. 202–203 ; see also Björnbo and Vogl, *Alkindi, Tideus und Pseudo-Euklid, Abhandl. z. Geschichte d. math. Wissen.*, XXXVI (Leipzig, 1912), p. 156, and the review of the same by A. Birkenmajer, *Bibliotheca Mathematica*, Vol. XIII, third series (1913), pp. 273–280.

prose of Omar Khayyam. The King of Saragossa, Jusuf al-Mutamin (reigned 1081–1085), was a devoted student of the mathematical sciences. The title of one of his works suggests the possibility that it was algebraical. For the study of law a knowledge of algebra seems to have been necessary, as various questions of inheritance were treated by this science, and even to-day in the great Mohammedan schools at Cairo and Mecca the study *aljebr w'al-muqabala* is considered of peculiar value to the prospective lawyer.

CHAPTER III

ROBERT OF CHESTER AND OTHER TRANSLATORS OF ARABIC INTO LATIN

WHEN towards the beginning of the twelfth century European scholars turned to Islam for light, the works of Mohammed ibn Musa came to occupy a prominent place in their studies. One of the first of these students was styled John of Seville,[1] or of Luna, or of Spain, whose name is attached to some manuscript copies of an adaptation of Al-Khowarizmi's arithmetic. Thus, the version, mentioned above, published by Prince Boncompagni, bears the title, *Joannis Hispalensis liber Algorismi de pratica arismetrice*, and in a sub-head the editing of the arithmetic is ascribed to John of Spain. This work contains also a very brief treatment of algebra, entitled *Exceptiones de libro, qui dicitur gleba mutabilia.*[2] *Res* is employed for the square of the unknown, and *radix* for the unknown, the usage being, in this respect, unique. The problems,

$$x^2 + 10\,x = 39,$$
$$3\,x + 4 = x^2,$$

occur as in Al-Khowarizmi, with a variation in the second type,

$$x^2 + 9 = 6\,x.$$

The authorship is in question, since some manuscripts ascribe the work to Gerard of Cremona, some to John of Spain, and others are anonymous. However, no doubt now exists that John of Spain was familiar with the arithmetic of Al-Khowarizmi, for Dominicus Gundisallinus, co-laborer with John in translating from the Arabic, mentions the *Liber algorismi* in the chapter on arithmetic of the *De divisione philosophiae.*[3] About 1133 A.D. Bishop Raimund of Toledo commissioned John of Spain to work with Gundisallinus on translations from the Arabic. John made the translation into Spanish, and this was put into Latin by Gun-

---

[1] Steinschneider, *Die Hebr. Uebers.*, p. 981, and note 82, p. 380.

[2] Unintelligible Latin forms for "algebra w' almucabala."

[3] Edited by L. Baur, *Beiträge z. Gesch. d. Philos. d. Mittelalters* (Vol. IV, Münster, 1903), p. 91.

disallinus. The probability is that the *De divisione philosophiae* is their joint production.[1] A somewhat similar method was pursued at first by Gerard of Cremona, collaborating with an Arab named Galippus or Galib.[2]

Adelard of Bath translated the astronomical tables which we have mentioned and possibly another astronomical work by Al-Khowarizmi.[3] His life is typical of the life of learned men of that period. Although born in England, he evidently went to France at an early age. There he studied at Tours and delivered lectures in Laon. At least seven years of his life were spent in study and travel in the East. Tarsus, Antioch, and Salerno are mentioned by him as cities which he visited. While no direct evidence is known that he studied in Spain, yet many of his works are based on Arabic documents transmitted to Europe through the Spanish schools at Toledo and Segovia. Learning was quite as international in that time as to-day.

Gerard of Cremona, too, desiring to find the works of Ptolemy, journeyed to Spain and there took up the study of the Arabic language in order to understand the Arabic version of Ptolemy, with the result that he devoted his life to translations from the Arabic. Included in an early list[4] of his translations is the algebra of Al-Khowarizmi, and it seems probable that the Latin version published by Libri[5] is from his hand. However, Boncompagni in his discussion of the life and works of Gerard of Cremona has published another mediæval adaptation of the algebra which is ascribed to Gerard. The words *res* and *census* for the unknown and its square, and also the title *aliabre et almuchabala*, are used by Gerard in his translation of Ababucri's *Book of the measurement of the earth and of solids*, as yet in manuscript.[6] Plato of Tivoli was also doubtless familiar with Mohammed ibn

[1] Steinschneider, *Die Hebr. Uebers.*, p. 981 and note 82, p. 380.

[2] Valentin Rose, *Ptolemaeus und die Schule von Toledo, Hermes*, Vol. VIII, pp. 332 ff.

[3] C. H. Haskins, *Adelard of Bath, The English Historical Review*, Vol. XXXVI (1911), pp. 491–498, and *Adelard of Bath and Henry Plantagenet, ibid.*, Vol. XXXVIII (1913), pp. 515–516.

[4] Boncompagni, *Della vita e delle opere di Gherardo Cremonese*, etc., *Atti dell' Accademia pontificia de' nuovi Lincei*, Vol. IV (1851), pp. 378–493.

[5] Libri, *Histoire des sciences mathématiques en Italie*, Vol. I (Paris, 1838), pp. 253–297; Björnbo, *Gerhard von Cremona's Uebersetzung von Alkwarizmi's Algebra und von Euklid's Elementen, Bibliotheca Mathematica*, Vol. VI (1905), third series, pp. 239–248.

[6] My statements are based upon the Paris MS. Latin 9335 and the Cambridge University Library MS. Mm. 2, 18, both of which contain the work in question.

Musa's works, for he mentions him as one of the Arabic math-ematicians.[1] Contemporary with these men was Robert of Chester.

The assertion was made by Curtze[2] that the *Liber embadorum*, *Book of Measures*, in Hebrew, by Abraham bar Chiyya Ha Nasi, known as Savasorda, was the first work to appear in Latin show-ing to the Western world how the solution of quadratic equations is accomplished. This statement is made on the basis of the date DX of the Hegira for the translation of the work made by Plato of Tivoli. It has recently been shown by Haskins,[3] on the basis of astronomical data in the work, that this date is undoubtedly a scribe's error for DXL, corresponding to 1145 A.D. Savasorda was approximately contemporary with his translator. Of the early Jewish writers many were familiar with the works of Al-Khowa-rizmi. Thus Abu Masar[4] cites the tables of our author while Abraham ben Esra[5] refers frequently to the same tables.

Three other Englishmen besides Adelard of Bath are known to have been students of Arabic mathematical science as taught in the schools of Spain in the twelfth century. The names and dates, as given by Wallis,[6] are : Adelard of Bath in 1130, Robertus Retinensis in 1140, William Shelley (de Conchis) in 1145, and Daniel Morley (Merlac) in 1180. This Robertus Retinensis[7] was also known as Robertus Ketenensis, de Ketene, Ostiensis,

[1] Favaro, *Intorno alla vita ed alle opere di Prosdocimo de' Beldomandi, Bullettino di bibliografia e di storia delle scienze matematiche e fisiche*, Vol. XII (Rome, 1879), pp. 1–74, 115–251. See pp. 122–123.

[2] M. Curtze, *Der Liber Embadorum des Savasorda in der Uebersetzung des Plato von Tivoli, Abhandl. z. Geschichte d. math. Wissen.*, Vol. XII (1902), p. 7.

[3] C. H. Haskins, *The Romanic Review*, Vol. II (1911), p. 2, note 5 ; *The English Historical Review*, Vol. XXVI (1911), p. 491, note 1.

[4] Steinschneider, *Zum Speculum des Albertus Magnus, Zeitschrift f. Mathematik und Physik*, Vol. XVI (1871), p. 376.

[5] Steinschneider, *Zur Geschichte der Uebersetzungen aus dem Indischen in's Arabische und ihres Einflusses auf die arabische Literatur, Zeitschrift der deutschen morgenländischen Gesellschaft*, Vol. XXIV (1870), pp. 339, 355 *et al.*

[6] Wallis, *A Treatise on Algebra, both historical and practical* (London, 1685), pp. 10–12.

[7] The *Dictionary of National Biography* includes two accounts of the life of Robert of Chester; Vol. X (London, 1887), p. 203, *Chester, Robert*, by A. M. Clerke; Vol. XLVIII (New York, 1896), pp. 362–364, *Robert the Englishman, Robert de Ketene*, or *Robert de Retines*, by T. A. Archer. Wüstenfeld, *Die Uebersetzungen arabischer Werke in das Lateinische, Abhandl. d. Königlichen Gesellschaft der Wissenschaften zu Göttingen*, Vol. XXII (1877), pp. 44–47 ; L. Leclerc, *Histoire de la médecine arabe* (Paris, 1876), Vol. II, pp. 380–387 ; Jourdain, *Recherches sur les anciennes traductions Latines d'Aristote* (Paris, 1843, revised edition), pp. 100–104 ; Thomas Wright, *Biographia Britannica Literaria, Anglo-Norman Period* (London, 1846), pp. 116–119.

Astensis, or Cestrensis, the final form being the most common.
Retinensis (Retenensis) has been somewhat doubtfully referred to
Reading, England, while Cestrensis certainly refers to Chester.
Similar peculiarities in the dual or multiple designation of writers
have been noted in connection with the name of John of Spain;
such variations seem to have been common in this period. Robert
of Chester, known to fame chiefly as the first translator of the
Qoran,[1] was doubtless educated in the well-known school located
at Chester. Of the more personal side of Robert's life we have
but scattered facts. His nationality is established not only by
his name and his return to England in 1150, but also by the direct
statement made by Peter the Venerable in a letter[2] of 1143 concerning the Qoran, addressed to Bernard of Clairvaux. Peter
states in this letter that Robert was then archdeacon of Pampe-
luna, in northern Spain. Hermann the Dalmatian, commonly
known as Hermannus secundus, but also spoken of as Scho-
lasticus, Sclavus, or Chaldæus, refers to Robert as his "special
and inseparable comrade, his peerless partner in every deed and
art." In the year 1141 Robert and Hermann were living in
Spain near the Ebro, studying the arts of astrology. There in
that year Peter the Venerable found them and "by entreaty and
a good price" induced them to take up studies in Mohammedan
religion and law, and also to translate the Qoran.[3]

[1] *Machumetis Saracenorum principis, eiusque successorum vitae, doctrina, ac ipse
Alcoran . . . cum doctiss. uiri Philippi Melancthonis praemonitione.* . . . Haec omnia
in unum uolumen redacta sunt, opera et studio Theodori Bibliandri (1550, place of publication, Basle, not given in book itself), Vol. I, pp. 213–223; a different edition was edited
by Wallis (Basle, 1643); possibly the first edition was printed at Basle, 1543. I have
used a copy of the edition of 1550, loaned to me from the John G. White Collection,
Cleveland Public Library, by the courtesy of the librarian.

[2] *Qoran*, 1550 edition, pp. 1–2; Migne, *Patrologia Latina*, Vol. 189 (Paris, 1890),
col. 649–652; see also col. 1073–1076, and col. 339.

[3] As there is some question as to the real translator of the Qoran it seems desirable to
add from the 1550 edition of the Qoran the evidence that the translation is due to Robert,
pp. 1–2: Epistola Domini Petri Abbatis, ad Dominum Bernhardum Claraeuallis Abbatem,
de translatione sua, qua fecit transferri ex Arabico in Latinum, sectam, siue haeresim,
Saracenorum. . . . Mitto uobis, charissime, nouam translationem nostram, contra pes-
simum nequam Machumet haeresim disputantem. Quae nuper dum in Hispaniis morarer
meo studio de Arabico uersa est in Latinam. Feci autem eam transferri a perito utriusque
linguae uiro magistro Petro Toletano. Sed quia lingua Latina non ei adeo familiaris uel
nota erat, ut Arabica, dedi ei coadiutorem doctum uirum dilectum filium et fratrem Petrum
notarium nostrum, reuerentiae uestrae, ut extimo, bene cognitum. Qui uerba Latina
impolite uel confuse plerumque ab eo prolata poliens et ordinans, epistolam, imo libellum
multis, ut credo, propter ignotarum rerum notitiam perutilem futurum perfecit. Sed et

In the prologue to his treatise[1] against the "sect" of the Saracens, Peter says: "Contuli ergo me ad peritos linguae Arabicae, . . . eis ad transferendum de lingua Arabica in Latinam perditi hominis originem, vitam, doctrinam, legemque ipsam quae Alchoran vocatur, tam prece quam pretio, persuasi. Et ut translationi fides plenissima non deesset, nec quidquam fraude aliqua nostrorum notitiae subtrahi posset, Christianis interpretibus etiam Sarracenum adjunxi. Christianorum interpretum nomina: Robertus Kecenensis, Armannus Dalmata, Petrus Toletanus; Saraceni Mahumeth nomen erat. Qui intima ipsa barbarae gentis armaria perscrutantes, volumen non parvum ex praedicta materia Latinis lectoribus ediderunt. Hoc anno illo factum est quo Hispanias adii, . . . qui annus fuit ab Incarnatione Domini 1141."

The letter to Bernard seems to be of date 1143, and that is the date, evidently, of the completion of Robert's translation, and so it is given at the end of the Qoran: "Illustri gloriosoque viro Petro cluniacensi Abbate praecipiente, suus Angligena, Robertus Retenensis librum istum transtulit Anno domini MCXLIII, anno Alexandri MCCCCIII, anno Alhigere DXXXVII, anno Persarum quingentesimo undecimo."

Robert and Hermann appear to have been associated in translating scientific works particularly along the lines of astrology from Arabic into Latin, as well as the Qoran.[2] In connection with these astrological treatises a mediæval reference,[3] probably contemporary, mentions Robert as "a man most learned in astrology." Peter the Venerable refers to Robert and Hermann as most acute and well-trained scholars, while Peter of Poitiers, in a letter[4] of unknown date addressed to Peter the Venerable, cites Robert

---

totam impiam sectam, uitamque nefarii hominis, ac legem, quam Alcoran, id est, collectaneum praeceptorum appellauit, sibique ab angelo Gabriele de coelo callatam miserrimis hominibus persuasit, nihilominus ex Arabico ad Latinitatem perduxi, interpretantibus scilicet uiris utriusque linguae peritis, Roberto Retenensi de Anglia, qui nunc Papilonensis ecclesiae archidiaconus est, Hermanno quoque Dalmata acutissimi et literati ingenii scholastico. Quos in Hispania circa Hiberum Astrologicae arti studentes inueni, eosque ad haec faciendum multo precio conduxi.

[1] *Patrologia Latina*, Vol. 189, col. 671.

[2] See the article by Björnbo, *Hermannus Dalmata als Uebersetzer astronomischer Arbeiten*, *Bibliotheca Mathematica*, Vol. VI, third series (1905), pp. 130-133 and p. 28 below.

[3] Steinschneider, *Die Europäischen Uebersetzungen aus dem Arabischen, Sitzungsberichte der Philosophisch-historischen Klasse der kaiserlichen Akademie der Wissenschaften*, Vol. 149 (Vienna, 1905), p. 72.

[4] *Patrologia Latina*, Vol. 189, col. 661.

as an authority on Mohammedan customs.   A further manuscript
reference [1] seems to indicate that in the year 1136 Robert was
studying in Barcelona with Plato of Tivoli, while Fabricius [2] states,
but upon what authority does not appear, that Robert travelled in
Italy, Greece, and Spain.   The time and the place of Robert's
death are equally uncertain.

To Peter the Venerable, Abbot of Cluny, who induced Robert
to undertake the translation of the Qoran, the latter addressed a
Saracenic chronicle, *Chronica mendosa et ridiculosa Saracenorum*,[3]
which was published in 1550 with the Qoran.   Other names have
been connected with this translation of the Qoran, but a study of
Peter's letters, and the introduction to Peter's treatise, " Against
the Sect of the Saracens," [4] shows that Robert and Hermann were
definitely requested to undertake the translation, which it appears
from Robert's preface was finally completed by Robert alone.
Possibly Peter of Toledo made an earlier,[5] unsatisfactory translation of the same work.   A Mohammedan by the name of
Mohammed was engaged by Peter the Venerable to scrutinize
the various treatises with a view to correcting errors due to
mistranslation.

Of Robert's works the version of the Qoran was completed in
1143.   In a prefatory letter he states that this task was regarded
by him only as a digression from his principal studies of astronomy and geometry,[6] but posterity knew of him through the

[1] Archer, *Dictionary of National Biography*, Vol. XLVIII (New York, 1896), p. 362 ;
Professor Haskins examined the MS. but found no evidence for this statement.

[2] *Bibliotheca med. et infim. Lat.* (Florence, 1858–1859), Vol. VI, p. 107.

[3] *Qoran*, 1550 edition, pp. 213–223.

[4] Migne, *Patrologia Latina*, Vol. 189, col. 659–720, Petri Venerabilis, Abbatis Clunia-
censis noni, adversus nefandam sectam Saracenorum libri duo.

[5] Since Peter the Venerable expressly refers to a new translation by the labors of
Robert and Hermann.   In the letter to Bernhard of Clairvaux Peter says : " Mitto vobis,
charissime, novam translationem nostram," which may mean, possibly, another document
on Arabic customs.

[6] *Machumetis Saracenorum principis, eiusque successorum vitae, doctrina, ac ipse
Alcoran* (Basle, 1550), pp. 7–8 : Praefatio Roberti translatoris ad Dominum Petrum
Abbatem Cluniacensem, in libro legis Saracenorum, quem Alcoran uocant, id est, Collec-
tionem praeceptorum quae Machumet pseudopropheta per angelum Gabrielem quasi de
coelo sibi missa confinxerit.   Begins, Domino suo Petro diuino instinctu Cluniacensi
abbati, Robertus Retenensis suorum minimus in Deo perfecté gaudere ; and ends, Istud
quidem tuam minimé latuit sapientiam, quae me compulit interim astronomiae geometriae-
que studium meum principale praetermittere.   Sed ne proemium fastidium generet, ipsi
finem impono, tibique coelesti, coelum omne penetranti, coeleste munus uoueo : quod
integritatem in se scientiae complectitur.   Quae secundum numerum, & proportionem

earlier work rather than through the sciences to which he dedicated himself in this letter.   Even the mathematician John Wallis, editor of one edition of this translation, did not connect the translator with the Robertus Retinensis who was mentioned, as we have seen, by Wallis in his *Algebra*.   An unpublished letter written by Robert, beginning *Cum jubendi religio*, is preserved in the Selden manuscript, sup. 31, in the Bodleian library.

Several sets of astronomical tables are included among the products of Robert's literary activity.   The *Canones in motibus coelestium corporum ad meridiem urbis Londoniarum in duos partes, prior 1149 ad fidem tabularum toletanarum Arzachelis, altera pro anno 1150 iuxta Albatem Haracensis*[1] evidently includes two sets of tables.   This reference to the methods of Al-Battani may account for the fact that Robert is sometimes credited with a translation of Al-Battani's tables.[2]   Wallis asserts that the tables accommodated to the meridian of London were adjusted to the year 1150 and in fact the text in MS. Savile 21, fol. 86[r], states the date as March first, 1150.   Reference is made also to a preceding work as accommodated to the meridian of Toledo for the year 1148 or 1149 and based on tables by Rabbi Abraham ibn Esra.[3]   Extracts, apparently from these tables of Robert, are found under the title, *De diuersitate annorum ex Roberto Cestrensi super tabulas toletanas*,[4] and this contains a reference to a sexagesimal multiplication table, evidently constructed by Robert, in which appear all products from 1 × 1 up to 60 × 60.

atque mensuram coelestes, circulos omnes, & eorum quantitates & ordines & habitudines, demum stellarum motus omnimodos, & earum effectus atque naturas, & huiusmodi caetera diligentissime diligentibus aperit, nunc probabilibus, nonnunquam necessariis argumentis innitens.

[1] Steinschneider, *Sitzungsber., loc. cit.*; MS. Savile 21, 63 –95[v]. Fol. 63[r], *Incipiunt canones in motibus coelestium corporum*, which begins, Quoniam cuiusque accionis quantitatem metitur celestium spacium, and proceeds with a discussion of various chronological systems.   Fol. 86[r], Incipit pars altera huius operis que videlicet ad meridiem urbis Londoniarum iuxta Albatem Haracensis sententiam per Robertum Cestrensem contexitur: Begins, Premissa uero electe facultatis . . . ; ends, Ergo iuxta hanc scienciam planetarum loca figura ponenda sunt.   See above, p. 17, note 1.

[2] See Nallino, *Al-Battānī opus astronomicum, Pubbl. del reale Osservatorio di Brera* (Milan), Vol. XL (1903), Introduction.

[3] In MS. Savile : *ebenza* (?)

[4] MS. Digby 17. Beg. fol. 156, Diuersi astronomi secundum diuersos annos tabulas faciunt et quidam secundum annos Alexandri seu grecorum, alii secundum ierdaguth (Yezdegerd) seu persarum.

In 1144 Robert completed the translation, entitled *De compositione alchemiae*, or *De re metallica*, by one Morienus Romanus.
This was published in Paris in 1546, and later by Manget.[1]
Another of the products of his literary activity, dated 1185 of the
Spanish Era (*circa* 1150 A.D.), was the work, *De compositione astrolabii*,[2] which is ascribed to Ptolemy. The place of composition is
given in one manuscript copy as London.[3] Like Adelard of Bath,
Robert seems also to have written a chemical work, dealing with
pigments and other associated topics, entitled *Liber metricus qui
dicitur Mappa claviculae*. This version, Greek in its origin, is in
verse.[4]

An astrological work, entitled *Judicia Alkindi astrologi* or
De judiciis astrorum, which was written by the great Arabic phi-

---

[1] J. J. Manget, *Bibliotheca Chemica Curiosa* (Geneva, 1702), Vol. I, pp. 509–519,
Liber De compositione Alchemiae, quem edidit Morienus Romanus, Calid Regi Aegyp-
torum ; quem Robertus Castrensis de Arabico in Latinum transtulit.

The prologue begins : Praefatio Castrensis. Legimus in Historiis veterum divinorum,
tres fuisse Philosophos, quorum unusquisque Hermes vocabatur. . . . Sed nos, licet in
nobis juvene sit ingenium, et latinitas permodica : hoc tamen tantum ac tam magnum opus
ad transferendum de Arabico in Latinum suscepimus. Vnde et de singulari gratia nobis
a Deo inter modernos collata, summas illi Deo vivo, qui trinus extat et unus, grates referimus.
Nomen autem meum in principio Prologi taceri non placuit, ne aliquis hunc nostrum
laborem sibi assumeret, et etiam ejus laudem et meritum sibi quasi proprium vendicaret.
Quid amplius ? humiliter omnes rogo et obsecro, ne quis nostrorum erga meum nomen
mentis livore (quod saepe a multis fieri consuevit) tabescat. Scit namque Deus omnium,
cui suam conferat gratiam : et spiritus ex gratia procedit, qui quos vult inspirat. Merito
igitur gaudere debemus quum omnium Creator et conditor omnibus quasi particularem
suam monstret divinitatem.

The treatise itself begins : Morieni Romani, Eremitae Hierosolymitani, Sermo. Omnes
Philosophiae partes, mens Hermetis divina plenarie attigit . . . and ends, Quod si quando
Alchymia confecta fuerit, ejus una pars inter novem partes argenti ponatur, quoniam totum
in aurum purissimum convertetur. Sic ergo Deus benedictus Amen, per cuncta seculorum
secula.

Explicit Liber Alchymiae de Arabico in Latinum translatus, anno millesimo centesimo
octuagesimo secundo, in mense Februarii et in ejus die undecimo.

[2] Steinschneider, *Zeitschrift f. Mathematik und Physik*, Vol. XVI, p. 393, with reference to the Vienna MS. 5311[8], *De compositione astrolabii universalis liber a Roberto Cas-
trensi translatus*, which begins : *Quoniam in mundi spera*, and ends : *ad altitudines
accipiendas; also to the Oxford MS. Cod. Canonic. Misc. 61[6], *Liber de officio astrolabii
secundum mag. Rob. Cestrensem*, in thirty-five chapters, which begins : 1. *De gradu solis
per diem et diei*, and ends : *et cetere ceteris per diametrum ut jam dictum est opponuntur*.
Revised after 1150, as it cites the tables of that year (Bodleian MS. Canon Misc. 61, f. 22[2] ;
examined by Professor Haskins).

[3] Steinschneider, *ibid.*, p. 393, refers to the Gonville and Caius College, Cambridge,
MS. 35[14], *Ptholomaei de compositione astrolabii, translatus a Rob. Cestiensi in civitate
London*; also MS. Digby 40, dates this " *Aera* 1185 *in civitate London.*"

[4] Steinschneider, *loc. cit.*, p. 72.

losopher Ya'qub ibn Ishaq ibn Sabbah al-Kindi (died about 874 A.D.), was translated by Robert of Chester and not by Robertus Anglicus in 1272 as usually[1] stated. This treatise in forty-five chapters is preceded by an introduction[2] which definitely establishes the authorship of the translation. In these prefatory remarks the writer addresses himself to "my friend Hermann, second to no astronomer of our time, of those who speak Latin," and further makes reference to the translation as being made at the request or wish of Hermann. As some manuscripts ascribe this work to Robert of Chester who was associated with Hermann the Dalmatian, the reference to Hermann would be almost conclusive, and the terminology of the introduction by its similarity to that of the Qoran introductory letter establishes the fact beyond question.

A rearrangement of the Al-Khowarizmian tables as translated by Adelard of Bath was made by Robert of Chester.[3] The basis

[1] Suter, *Die Mathematiker und Astronomen der Araber und ihre Werke, loc. cit.*, p. 26; Steinschneider, *Die europäischen Uebersetzungen aus dem Arabischen, loc. cit.*, p. 66, *et al.*

[2] Introduction, for which I am indebted to the courtesy of Professor C. H. Haskins, following MS. Ashmole, 369, fol. 85ᵃ. The heading and *explicit* are from the Cotton MS. App. vi. Numerous other manuscripts of the same are found in European libraries.

Incipiunt iudicia Alkindi astrologi, Rodberti de Ketene translatio.

Quamquam post Euclidem Theodosii cosmometrie libroque proportionum libentius insudarem, unde commodior ad almaiesti quo praecipuum nostrum aspirat studium pateret accessus, tamen ne per meam segnitiem nostra surdesceret amicitia, vestris nutibus nil preter equum postulantibus, mi Hermanne, nulli latinorum huius nostri temporis astronomico sedere (sedem?) penitus parare paratus, eum quem commodissimum et veracissimum inter astrologos indicem vestra quam sepe notauit diligencia, voto vestro seruiens transtuli non minus amicitie quam pericie facultatibus innisus. In quo tum nobis tum ceteris huius scientie studiosis placere plurimum studens, enodato verborum vultu rerum seriem et effectum atque summam stellarium effectuum pronosticationisque quorumlibet eventuum latine brevitati diligenter inclusi. Cuius examen vestram manum postremo postulans non indigne vobis laudis meritum si quod adsit communiter autem fructus pariat, mihique non segne res arduas aggrediendi calcar adhibeat, si nostri laboris munus amplexu fauoris elucescat. Sed ne proemium lectori tedium lectionique moram faciat vel afferat, illius prolixitate supersedendo, rem propositam secundum nature tramitem a toto generalique natis exordiis texamus, (P)rius tamen libri totius capitulis enunciatis ad rerum evidentiam suorumque locorum repertum facilem. *Explicit prohemium. Incipiunt libri capitula.*

Primum igitur capitulum, zodiaci diuisiones, earumque proprietates, tam naturales quam accidentales generaliter complectitur. . . .

The text proper begins (Ashmole MS.): Circulus itaque spericus cuius atque terre centrum est . . . and ends: sequitur in proximo.

[3] Haskins, *The English Historical Review*, XXVI, footnote on p. 498: Madrid MS. (no. 10016) of the translation of the Al-Khowarizmian tables, which has this heading: *Incipit liber ezeig id est chanonum Alghoarizmi per Adelardum bathoniensem ex arabico sumptus et per Rodbertum cestrensem ordine digestus!* Suter, *Die astron. Tafeln des al-Khwarizmi in der Bearbeitung des Maslama ibn Aḥmed al-Madjriṭi und der lat. Uebers. des Athelard von Bath auf Grund der Vorarbeiten von A. Björnbo † und R. Besthorn, loc. cit.*

of this revision was doubtless a translation of the same tables by Hermann, who refers to his own translation in his astrological work, *Introductorium in astronomiam Albumasaris abalachi, octo continens libros partiales.*[1]   This work is dedicated to Robert; the similarity of the phraseology with that of the prefaces by Robert bears witness to the intimacy of their literary labors.   Also, in the prologue to Hermann's translation of Ptolemy's *Planisphere,*[2] Robert is mentioned as an associate of Hermann and again in the text.   This work on the Planisphere has been incorrectly ascribed in the printed edition of 1537 and elsewhere to Hermann's pupil, Rudolph of Bruges.[3]

[1] H. Suter, in a letter (Aug. 13, 1914) to the author.

The *Introductorium* was printed at Augsburg in 1489, and other editions followed.   I am indebted to Professor Haskins for a transcription of the preface, from which the following passage is taken: Que cum ego prolixitatis exosus et quasi minus contentia: cum et hunc morem latinis cognoscerem preterire volens animo ipso potius tractatum exordiri pararem.   Tu mihi studiorum olim specialis atque inseperabilis comes, rerumque et actuum per omnia consors unice, mi Rodberte, si memores obviasti dicens: Quanquam equidem nec tibi pro amore tuo, mi Hermanne, nec ulli consulto aliene lingue interpreti in rerum translationibus. . . .

Steinschneider, *Ueber die Mondstationen (Naxatra) und das Buch Arcandum, Zeitschrift d. deutschen morgenländischen Gesellschaft,* Vol. XVIII (1864), pp. 170–172, and *Die Hebräischen Uebersetzungen,* pp. 568–570, and in *Die europäischen Uebersetzungen, loc. cit.,* p. 34, demonstrated from the introduction that Hermann was the author.   This view was accepted by Björnbo, *Hermannus Dalmata als Uebersetzer astronomischer Arbeiten, Bibliotheca Mathematica,* third series, Vol. IV (1903), pp. 130–133.

[2] J. L. Heiberg, *Claudii Ptolemaei Opera quae exstant omnia,* Vol. II, *Opera astronomica minora,* pp. clxxxiii–clxxxvi: "tuam itaque uirtutem quasi propositum intuentes speculum ego et unicus atque illustris socius Rebertus Ketenensis nequitie dispicere, licet plurimum possit, perpetuum habemus propositum, cum, ut Tullius meminit, misera sit fortuna, cui nemo inuideat."

In the text proper Hermann inserts the following note: "quem locum a Ptolomeo minus diligenter perspectum cum Albateni miratur et Alchoarismus, quorum hunc quidem ope nostra Latium habet, illius uero comodissima translatione studiosissimi Roberti mei industria Latine orationis thesaurum accumulat, nos discutiendi ueri in libro nostro de circulis rationem damus."

[3] After this chapter was in type a noteworthy article, *The reception of Arabic science in England,* by Professor C. H. Haskins, appeared in *The English Historical Review,* Vol. XXX (1915), pp. 56–69.   The works of Robert of Chester are carefully considered.   Fortunately, through the courtesy of Professor Haskins, I was enabled to incorporate in advance the results which bear directly upon our discussion.

CHAPTER IV

The Influence of Al-Khowarizmi's Algebra upon the Development of Mathematics

By the translators of Arabic lore we are brought from Islam to Christendom. Mathematical science in Europe was more vitally influenced by Mohammed ibn Musa than by any other writer from the time of the Greeks to Regiomontanus (1436–1476). Through his arithmetic, presenting the Hindu art of reckoning, he revolutionized the common processes of calculation and through his algebra he laid the foundation for modern analysis. Evidence of the influence of the great Arab is presented by the relatively large number of translations and adaptations of his various mathematical works which appeared before the invention of printing. Undoubtedly the earliest translation of the Arabic algebra, although not the most widely used, was that made by Robert of Chester. Probably the version published by Libri appeared shortly afterwards for, as we have mentioned, Gerard of Cremona employs the terminology of that version in algebraic work. Roger Bacon (1214–1294), too, has occasion [1] to mention the algebra, as well as the arithmetic, and uses terms not found in Robert of Chester's version. Bacon shows that he had but superficial familiarity with the subject, for he made incorrect statements about the fundamental elements of the algebra. Similarly, Vincent de Beauvais (about 1275) in his encyclopædic work, *Speculum Principale*,[2] refers under arithmetic to the book, *qui apud Arabes mahalehe dicitur*. Albertus Magnus (1193–1280) mentions the tables of Al-Khowarizmi.[3]

Even earlier than Roger Bacon is Leonard of Pisa, whose monumental *Liber abaci* contains a chapter involving the title, *Aljebra*

---

[1] In his unfinished *Scriptum principale*. See J. H. Bridges, *The "Opus Majus" of Roger Bacon* (London, 1900), Vol. I, p. lvii; D. E. Smith, *The place of Robert Bacon in the history of mathematics*, in *Roger Bacon, Essays*, collected and edited by A. G. Little (Oxford, 1914), p, 177.

[2] Liber xviii, Cap. V, *De Arithmetica*; Liber xviii, Cap. ix, is entitled, *De Computo et algorismo*, and takes up representation of numbers by the Hindu numerals.

[3] Steinschneider, *Zum Speculum astronomicum des Albertus Magnus*, Zeitschrift f. *Mathematik und Physik*, Vol. XVI, pp. 375–376, and in *Zeitschrift d. deutschen morgenl. Gesellschaft*, Vol. XXV, pp. 404–405.

et almuchabala.[1] The first draft of this work was written in 1202, and in 1228 a revised and enlarged version appeared, dedicated to Michael Scotus. Woepcke has shown that Leonard drew many of his problems from Al-Khowarizmi,[2] but some of these may have come indirectly through Abu Kamil, from whom, as I have shown,[3] Leonard took many of his algebraic problems. In the manuscripts of the Italian's treatise the only mention of Al-Khowarizmi is in the margin, simply *Maumet*, at the beginning of the section dealing with algebra; but the term *algorismus* occurs for arithmetic.[4]

In the century following Leonard of Pisa, another Italian mathematician, William of Luna, is reputed to have put Al-Khowarizmi's algebra into the Italian language. Raffaello di Giovanni Canacci, a Florentine citizen of the fifteenth century, states in an Italian work on algebra,[5] as yet in manuscript, that William had translated the rules of algebra out of Arabic into " our language." Reference to his work is also made by at least three writers of the sixteenth century, the Florentine Francesco Ghaligai, the Spaniard Marco Aurel, and another Spaniard Antich Rocha of Gerona.[6] An Italian manuscript of 1464 in the library of George A. Plimpton, Esq., of New York, does contain an Italian version of the Algebra of Al-Khowarizmi in which reference is made to William of Luna as a translator of algebra. The possibility is that we have here the version of William. The writer of the manuscript is not known, but he explicitly states that he bases his treatise on the labours of numerous predecessors in this field. One chapter, as I have shown in a recent study of this manuscript,[7] deals with the algebra of an unknown Maestro Biagio (died 1340) and to Leonard of Pisa another section is devoted. The writer purposed also to deal with the works of a "subtle Maestro Antonio," doubtless Antonio Mazzinghi da Peretola, who wrote a treatise on algebra called *il fioretto*, and a " Maestro

---

[1] *Scritti di Leonardo Pisano* (Vol. I, Rome, 1857; Vol. II, Rome, 1862), Vol. I, p. 406.

[2] *Extrait du Fakhri*, p. 29.

[3] *The Algebra of Abu Kamil*, loc. cit.

[4] *Scritti*, Vol. I, p. 1.

[5] Codex Palat. 567, Biblioteca Nazionale, Florence.

[6] Ghaligai, *Pratica d'arithmetica* (Florence, 1552); Aurel, *Libro primero, de arithmetica algebratica* (Valencia, 1552); Rocha, *Arithmetica* (Barcelona, 1565).

[7] *An Italian Algebra of the fifteenth century*, *Bibliotheca Mathematica*, third series, Vol. XI, pp. 209-219.

Giouanni," but either this manuscript is incomplete or the plan was not carried out. Prominent also in the discussions is an Augustinian monk, Gratia de Castellani (about 1340), famed as a theologian. The relatively large number of names of men who had evidently attained something more than local repute in the study of algebra shows the place which it had reached in instruction.

Another prominent writer of the fourteenth century, Johannes de Muris, included a discussion of algebra in the third book of his popular *Quadripartitum numerorum*.[1] Of this section of the work of John of Meurs I have recently made a study,[2] showing that he drew extensively from Leonard of Pisa and from Al-Khowarizmi, thus continuing the Arabic influence. Regiomontanus included the work in a list of important early works on mathematics, and further Regiomontanus refers to algebra as the *ars rei et census*. This corresponds to a line of the *Quadripartitum*:

" Que tamen ars minor est quam sit de censibus et rei."

Later the expression, *Arte magiore*,[3] or *Ars mayor*,[4] or *Ars magna*,[5] was used for algebra, tracing back to this passage here given, in which *ars minor* refers to arithmetic as opposed to algebra. Adam Riese presents the problem,[6] $x^2 + 21 = 10\,x$, as being found in the eleventh chapter of the third book of the *Quadripartitum*; we have mentioned that this problem is one of the type problems found in Al-Khowarizmi's algebra. Another French author who gives an adaptation in Latin of the Arabic algebra is Rollandus, Canon of St. Chapelle. At the command of John, Duke of Lancaster, Rollandus wrote in the year 1424 a compendium of mathematics; the labor of composition was considerably lightened by making large extracts from the *Quadripartitum*, including most of the arithmetic and algebra. A summary of the contents of the

---

[1] Two chapters of the second book, dealing with arithmetic, were published by A. Nagl, *Das Quadripartitum etc.*, *Abhandl. z. Geschichte der mathem. Wissenschaften*, Vol. V (Leipzig, 1890), pp. 137–146.

[2] *The " Quadripartitum numerorum " of John of Meurs, Bibliotheca Mathematica*, third series, Vol. XIII, pp. 99–114.

[3] Paciuolo, *Summa d'arithmetica* (Venice, 1494).

[4] In Aurel's *Libro primero, de arithmetica algebratica* (Valencia, 1552).

[5] Cardan, *Ars Magna* (Nuremberg, 1545).

[6] B, Berlet, *Adam Riese, sein Leben, seine Rechenbücher und seine Art zu rechnen, Die Coss von Adam Riese* (Leipzig, 1892), p. 37.

manuscript is given in the *Rara Arithmetica*,[1] but the somewhat extensive treatment of algebra is not mentioned.

The first work in the German language on algebra was an excerpt from Al-Khowarizmi which begins:[2] "Mohammed in the book of algebra and almucabala has spoken these words 'census, radix (root), and number.'" This is followed by two problems from the text. The manuscript which contains this brief discussion, of date 1461, is now in Munich, having been moved from the Benedictine Abbey of St. Emmeran. The first treatment in the English language appears to be that by Robert Recorde, *The Whetstone of Witte*, which was published in 1557. This work which, as I have elsewhere shown,[3] does not display any marked originality on the part of Recorde, introduced our present symbol of equality, =, and contributed to the study of algebra in England by presenting the material in the mother tongue.

Regiomontanus seems to have been familiar with Al-Khowarizmi's work, for he not only refers to the art of thing and square (*ars rei et census*), but also uses certain technical expressions, *restaurare defectus*, for example, similar to those in the algebra. A manuscript copy[4] of Mohammed ibn Musa's algebra in Mr. Plimpton's collection shows astonishing similarity to the handwriting and abbreviations of Regiomontanus as well as to the form of equation used by the great German. Furthermore, some of the problems given in this manuscript, which are not part of Al-Khowarizmi's text, are discussed by Regiomontanus in his correspondence with Cardinal Blanchinus.[5] We must suppose him to have been familiar with this text if not actually, as we suspect, the transcriber of this copy. Regiomontanus was twenty years of age when this manuscript was written (1456), and we know that he did transcribe numerous mathematical and astronomical works of historical importance. Algebra was em-

---

[1] *Rara*, 446–447.

[2] Gerhardt, *Zur Geschichte der Algebra in Deutschland, Monatsbericht der Königl. Akad. d. Wissensch. zu Berlin*, 1870, pp. 141–153.

[3] Karpinski, *The whetstone of witte, Bibliotheca Mathematica*, third series, Vol. XIII, pp. 223–228.

[4] *Rara Arithmetica*, 454–456.

[5] M. Curtze, *Der Briefwechsel Regiomontan's mit Giovanni Bianchini, Jacob von Speier und Christian Roder, Abhandl. z. Gesch. d. math. Wissen.*, Vol. XII (Leipzig, 1902), problem 6, p. 219.

ployed by Regiomontanus in his trigonometry[1] in the solution of problems.

Regiomontanus has been cited by Nesselmann[2] as an illustration of one who employed rhetorical algebra as opposed to syncopated or symbolical. Later writers have followed Nesselmann in the assertion that Regiomontanus used rhetorical algebra, but, whereas the statement in Nesselmann is correct in so far as the illustration which he gives is concerned, the assumption that this was the general practice of the great Teuton is an error. In fact, his correspondence with Cardinal Blanchinus shows that he had a form of equation little inferior to ours. The + sign which he uses is a ligature for *et*, the minus sign a ligature for *minus*, and for an equality sign he uses a single straight line. Further he has separate symbols for the various powers of the unknown up to the cube, so that Regiomontanus approached modern forms more closely than most mathematicians even of the sixteenth century. This attempted division of the history of algebra into rhetorical, syncopated, and symbolic periods is an excellent illustration of a plausible and taking theory, in historical matters, which lacks only the first essential for such a theory; namely, historical evidence. Development in mathematics, as in art and literature, does not proceed in a logical manner, but rather in waves advancing and receding, and yet withal constantly advancing.

We have mentioned the Hebrew translation of the algebra of Abu Kamil, which was made by Mordechai Finzi (about 1475 A.D.) of Mantua. Another treatise on algebra, in Hebrew, dedicated to Finzi, was written by Simon Motot.[3] As the words *cosa* and *censo* are mentioned by Motot as being found in the works of Christian authors with which he was familiar the Italian source of his information is established, although the particular writers in question are not known. As we have above indicated, the Italians were sufficiently active in this science, so that many Italian works on algebra were available, in manuscript, at this time.

In the summer of 1486 Johann Widmann of Eger is known to

[1] Regiomontanus, *De triangulis libri quinque* (Nuremberg, 1533), problem 12, p. 51.

[2] Nesselmann, *Die Algebra der Griechen* (Berlin, 1842), p. 303.

[3] G. Sacerdote, *Le livre de l'algèbre et le problème des asymptotes de Simon Motot*, *Revue des études juives*, 1893-1894.

have lectured on algebra in the university at Leipzig, and the fee for the course was set extraordinarily high, being two florins.[1] Widmann was in possession of the Dresden manuscript,[2] which contains Robert of Chester's version of the algebra of Al-Khowarizmi, and himself added certain algebraic problems to another part of the same manuscript dealing with algebra. The Arab Al-Kalasadi,[3] contemporary with Widmann, wrote also on similar topics, and although he does not cite Al-Khowarizmi, yet he continues the old order of the six types of quadratic equations.

Adam Riese wrote in 1524 a work on algebra entitled, *Die Coss*, which contains, as we have noted, the problem

$$x^2 + 21 = 10\,x.$$

Riese refers to "that most celebrated Arabic master Algebra, learned in number, whose like in computation there never was, and hardly will any one exceed him." He refers also to the book, "named gebra and almucabola," by this mythical Algebra. A reference to Algum is also doubtless to Al-Khowarizmi. Several contemporaries, students of the Coss, are mentioned[4] by Adam Riese, and included among these is Grammateus,[5] also known as Schreiber or Scriptor, to whom is credited the first algebra in print in the German language. Interesting is Riese's note that Hans Conrad, to whom he frequently refers, paid the mathematician Andreas Alexander one florin in gold, to be taught how to solve certain types of problems by the Coss. This title is from the Italian *cosa* (Latin *res*, Arabic *shaï*) and connects with the use by Al-Khowarizmi, and subsequent Arabs, of the word *shaï*, meaning *thing*, for the first power of the unknown. For centuries the title continued in circulation in Germany, and even in English appeared in the form, "the arte of cosslike nombers."[6]

Luca Paciuolo, otherwise Luca de Burgo San Sepulchro, to whom is credited the first printed work on algebra, was evidently

---

[1] Wappler, *Zur Geschichte der deutschen Algebra in 15. Jahrhundert*, in *Programm*, Zwickau, 1887, pp. 9–10; also in *Zeitschrift f. Math. und Physik, Hist.-lit. Abtheil.*, Vol. 45, 1900.

[2] Codex Dresden C. 80.

[3] Woepcke, *Atti dell' accad. pont. de' nuovi Lincei*, Vol. XII (Rome, 1859), 230–275, 399–438.

[4] Berlet, *loc. cit.*, pp. 33, 34, 36, and 62 for references.

[5] . . . *Rechenbüchlin*, including *Etlichen Regeln Cosse*, written 1518 and published 1521.

[6] Recorde, *loc. cit.*

influenced by Al-Khowarizmi.   Paciuolo gives[1] the equation

$$x^2 + 10x = 39$$

and presents the geometrical explanation as given by the Arab. In certain others of the early printed algebras the fundamental or type equations as given by Al-Khowarizmi do not appear. Of such are the works by Grammateus appearing in 1518, by Christian Rudolph in 1525,[2] and by Estienne de la Roche in 1520,[3] whose work is known to have been a plagiarism of the *Triparty* by Nicolas Chuquet (1484).[4] However, many other writers did continue the type problems of the first systematic treatise. Thus Elia Misrachi[5] in an arithmetic which appeared in Constantinople in 1534, eight years after the author's death, devotes a section to algebra, and this is to a large extent an adaptation of the *Algebra and Almucabala*. In the work by Perez de Moya (1562)[6] and in the arithmetic of 1539 by Cardan we come upon the type equations. Mennher de Kempten, a Dutch mathematician, states that Algorithmo was the first writer on algebra. Ghaligai, the Italian, and the Spaniard Pedro Nuñez follow the peculiar order of equations found in Al-Khowarizmi. In these and other ways we might trace through the centuries the persistent influence, direct and indirect, of our Arabic author, but that is beside our present purpose.

Among the writers who made a serious study of Robert of Chester's translation we must place Johann Scheybl (1494–1570), who was professor of mathematics at Tübingen from about 1550 to the time of his death. He was the author of an algebra which appeared in two editions in Paris, 1551 and 1552. This treatment of algebra was first published in 1550 by Scheybl, prefixed to his Greek and Latin edition of the first six books of Euclid. Scheybl

---

[1] *Summa de arithmetica* (Venice, 1494), fol. 146 rec.

[2] I have not seen a copy of this edition. My remark is based upon *Die Coss Christoffs Rudolffs* (Königsberg, 1553) by Stifel. From certain notes about the history of the terminology and the words *dragma*, *res* and *substantia*, and the like, it appears that Stifel had seen a copy of Robert of Chester's version.

[3] Smith, *Rara Arithmetica*, p. 128.

[4] A. Marre, *Notice sur Nicolas Chuquet et son Triparty en la science des nombres*, Boncompagni's *Bulletino di bibliografia e di storia delle scienze matematiche e fisiche*, Vol. XIII (1880), pp. 555–592; the text of the *Triparty*, same volume, pp. 593–659 and pp. 693–814; the third section from 736–814 deals with quadratic equations and problems.

[5] G. Wertheim, *Die Arithmetic des Elia Misrachi* (Braunschweig, 1896), pp. 54–59.

[6] *Arithmetica practica, y speculativa* (Salamanca, 1562).

prepared for publication the Latin version of Al-Khowarizmi's algebra as translated by Robert of Chester. His manuscript copy is now in the Columbia University library. The title page [1] reads, in translation: 'A brief and clear exposition of the rules of algebra by Johann Scheybl, Professor of Euclid in the famous University of Tübingen. To this is added the work, *On given numbers*, by that most excellent mathematician Jordanus. Furthermore there is presented the book containing the demonstrations of the rules of algebraic equations, written some time ago in Arabic. All of these are now published for the first time by the above-mentioned Scheybl. These are corrected as far as possible and illustrated by appropriate and useful examples.'

The algebra contained in this manuscript is not the same as the published work mentioned above. However, the method of treatment is not materially different. The work by Jordanus Nemorarius, entitled *De numeris datis*, dates from the early part of the thirteenth century. The importance of the work, chiefly with respect to the development of algebra, is well attested by the fact that Regiomontanus and Maurolycus [2] both planned to publish the work, although neither carried the plan to completion. The work was published in 1879 by Treutlein.[3] The Scheybl version contains the complete list of 113 propositions to which Chasles made reference in 1841. These are divided into four books containing respectively 30, 26, 22 and 35 propositions. Scheybl adds solutions by the rules of algebra [4] in which he employs the same algebraic notation as in his published algebra, but he does not give the complete text of the work of Jordanus. The other two works in the manuscript are presented in this monograph. The manuscript was carefully prepared, but, for some reason which we do not know, the publication was not accomplished.

[1] Breuis ac dilucida regularum Algebrae descriptio, autore Joanne Scheubelio, in inclyta Tubingensi academia Euclidis professore ordinario.

Huic accedit liber consumatissimi mathematici Jordani, de datis.

Liber praeterea, continens demonstrationes aequationum regularum Algebrae, Arabice olim conscriptus.

Quae (*corr.* Qui) ambo ab eodem Scheubelio nunc primum, quantum fieri potuit, emendato (*corr.* emendati) in lucem aedita, et aptissimis atque utilibus exemplis illustrata sunt.

[2] Treutlein, *Abhandlungen zur Gesch. d. Math.*, Vol. II (Leipzig, 1879), pp. 127–128.

[3] Treutlein, *loc. cit.*, pp. 127–166; corrected by Curtze, *Zeitschrift f. Mathematik und Physik*, Vol. XXXVI (1891), pp. 1–23, 41–63, 81–95, and 121–138.

[4] Frequently adding, *Sequitur solutio ex regula Algebrae*.

PLATE II.

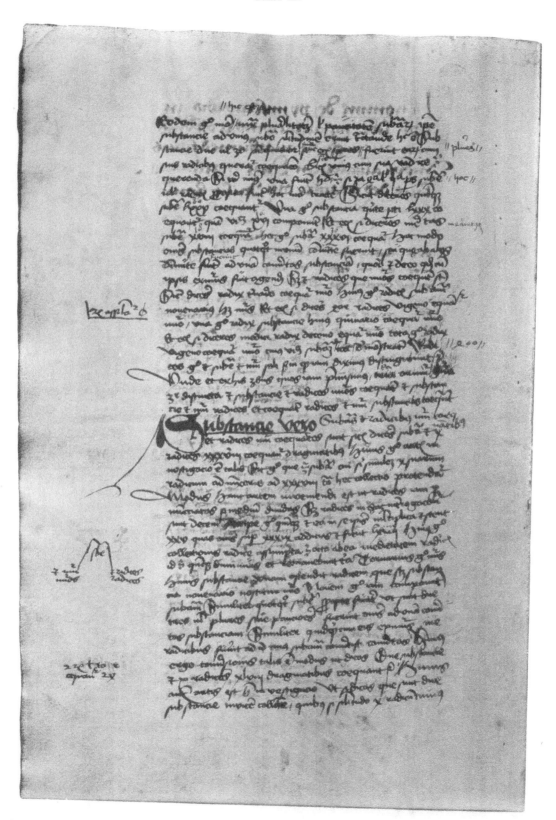

CODEX DRESDENSIS C. 80, Fol. 340ᵇ

A few words about the life of Scheybl[1] may be of interest.   His student days include an early stay at the University of Vienna, made famous in mathematical studies by Peurbach and Regiomontanus.   In 1532 Scheybl matriculated at Tübingen, which was a stronghold of Protestantism, and in 1535 he was a student there. In 1540 he was Magister in Tübingen, and four years later Docent in mathematics.   After another period of about five years we find him Professor (*ordinarius*) of Euclid, and in 1555 Professor of Euclid and Arithmetic.   How little some aspects of university life have changed during four centuries is shown by the fact that Scheybl twice, in 1551 and 1562, requested of the university authorities an increase of salary in order that he might pay his debts and obtain the necessaries of life.   In addition to the treatises on geometry and algebra which have been mentioned, Scheybl published other works on geometry and arithmetic.

As late as the end of the sixteenth century an able mathematician, Adrien Romain, deemed the algebra of Al-Khowarizmi sufficiently worthy of serious study to justify him in publishing a commentary on the work.   The version upon which Romain based his study is that of Robert of Chester.   In the course of his commentary he gives small portions of this translation.   At the time of writing this, according to Bosmans[2] in 1598 or 1599, Romain was teaching in Würzburg.   His student and teaching life covered periods of residence in Germany, Italy, and Poland, besides his native Louvain where he studied and taught.

[1] H. Staigmüller, *Johannes Scheubel, ein deutscher Algebraiker des XVI Jahrhunderts Abhand. zur Gesch. d. math. Wissenschaften*, Vol. IX, pp. 431–469.

[2] H. Bosmans, S.J., *Le fragment du commentaire d'Adrien Romain sur l'algèbre de Mahumed ben Musa el-Chowarezmi, Annales de la Société scientifique de Bruxelles*, Vol. XXX (1906).

CHAPTER V

THE ARABIC TEXT AND THE TRANSLATIONS OF AL-KHOWARIZMI'S ALGEBRA

THE Arabic text of Al-Khowarizmi's algebra, together with an English translation, was published by Frederick Rosen in 1831. This excellent[1] work is unfortunately out of print, and so is not available to most students of mathematics. The translation is made with care and intelligence, but not literally. Thus the Arabic invocation to the Deity is frequently omitted, just as it often is by modern translators.

The translation of the Algebra into Latin was made not only by Robert of Chester, but also, as we have indicated, by some other student of Arabic science who lived about the same time as Robert. This Latin version, as found in two Paris manuscripts, was published by Libri;[2] Björnbo believed that he had established[3] this to be the work of Gerard of Cremona, which indeed is probable. A list of the numerous translations due to Gerard was made soon after Gerard's death by some friend and admirer, and the list was published by Boncompagni.[4] Included among the titles is the algebra, *Liber alchoarismi de iebra et almucabula tractatus I.* However, the question is somewhat complicated by the fact that a mediæval adaptation of the algebra which was published by Boncompagni[5] bears the name of Gerard of Cremona. The text of this version does not follow the Arabic at all closely, and there is little reason for considering it as a direct translation. Probably the meaning of the title[6] is that the text of this version is based upon Gerard's translation.

---

[1] Portions of the Arabic text and translation have been examined by Professor W. H. Worrell, to whose courtesy I am indebted for the information about the character of the translation.

[2] *Histoire des sciences mathématiques en Italie* (Paris, 1838), Vol. I, pp. 253–297.

[3] *Gerhard von Cremona's Übersetzung von Alkwarizmi's Algebra und von Euklid's Elementen, Bibliotheca Mathematica*, third series, Vol. VI (1905), pp. 239–248.

[4] *Della vita e delle opere di Gherardo Cremonese etc., Atti dell' Accademia de' nuovi Lincei*, Vol. IV (1851), pp. 4–7.

[5] *Loc. cit.*, pp. 28–51.

[6] *Loc. cit.*, p. 28: Incipit liber qui secundum Arabes vocatur algebra et almucabala, et apud nos liber restauracionis nominatur, et fuit translatus a magistro Giurardo cremonense in toleto de arabico in latinum.

The Libri text varies essentially in phraseology and construction from that by Robert. The Arabic is closely followed up to the long list of problems, "Various Questions." Even here all the problems with the exception of two[1] are given in the Latin by Gerard, but not absolutely in the order in which they occur in the Arabic. The slight changes in the sequence of problems may well have been the fault of the particular Arabic manuscript which Gerard used, if it is not due to some transcriber of Gerard's work. One problem which is omitted is not very clear in the Arabic, but the second omission is a problem of the same type as others which are given. Some other slight omissions are made in the Latin text, and the longest of these corresponds to the passage in our text p. 84, line 25 to p. 86, line 2. Another omission in the Libri text corresponds to our text, page 74, line 25, *quod . . . reperies.* The Libri text also frequently omits the common invocation to the Deity which is so often interjected by Arabic writers.

The Latin translation by Robert of Chester is not as faithful nor as correct as the text ascribed to Gerard of Cremona, published by Libri. Omissions, transpositions, and additions to the text are so numerous that it does not seem desirable to list them all. No evidence exists, however, that Robert's text is based upon another Arabic original than that of the Libri text. The text proper, as opposed to the illustrative problems, follows the general lines of the Arabic original. The longest omission is the section dealing largely with the operations upon the square root of 200, which is illustrated, in the Arabic and in the Libri text, by geometrical figures with corresponding demonstrations.[2]

A sentence is left out on page 98 of our text, line 6, after the word *aequiparatur*. This sentence Rosen translates, 'Compute in this manner every multiplication of the roots, whether the multiplication be more or less than two.' Lines 9–11, *Natura . . .*

---

[1] Rosen's translation, p. 48, line 15 to p. 50, line 5, and p. 53, lines 12–20. Neither of these problems is given by Robert of Chester, nor does either appear in the Boncompagni version. The first problem reads: "If some one say: 'I have purchased two measures of wheat or barley, each of them at a certain price; I afterwards added the expenses, and the sum was equal to the difference of the two prices, added to the difference of the measures.'" The second reads: "Three-fourths of the fifth of a square are equal to four-fifths of its root."

[2] Rosen, *loc. cit.*, p. 27, lines 5–18, and p. 31, line 11, to the bottom of p. 34; Libri, *loc. cit.*, p. 269, lines 2–12, and p. 271, line 16–p. 274, line 14. The Libri version omits the statement of one problem, as stated by Rosen, p. 27, lines 14–16, but the geometrical explanation is complete.

fractionibus, on the same page of our text, seems to be an addition by Robert. The introduction of the passage, 'On Mercantile Transactions,' pp. 120–124, is not at all carefully translated by Robert, who retains poor transliterations of four technical expressions used in the Arabic. The four expressions in question refer to the four terms of a proportion in which when three are given the fourth is determined. If a given quantity of goods is sold at a fixed or set price, then the price of any other quantity of the same goods, or the amount of goods to be obtained for a given sum of money, is determined by a proportion in which the three given quantities enter. The unit of measure, or quantity sold at a fixed price, is termed by Robert *Almusarar*, instead of *al-musa-ʿir*, and the fixed price *Alszarar*, instead of *al-siʿr;* the quantity of goods desired is *Almuthemen*, instead of *al-muthamman* and the amount to be expended for goods is termed *Althemen*, instead of *al-thaman*. *Magul*, which is used by Robert for the unknown term in a proportion, would be in modern transliteration *al-maqul*.

Robert of Chester does not present the complete list of problems which occur in the Arabic text of Al-Khowarizmi's algebra, but only a selection of about one-half of the total number. Upon what basis this selection was made does not appear, except that typical problems are chosen, and the repetitions which are found in the Arabic and the Libri text are eliminated. In the footnotes to our English version we have indicated the problems which have been omitted by our author.

The translation of the text and solutions of the problems which are given present peculiarities entirely similar to those which have been noted in the preceding discussion of the Latin text by Robert. A noteworthy omission is made both by Robert of Chester and by the translator of the version published by Libri. This concerns the fifth problem of the set of six which illustrate in order each of the six types of quadratic equations. After the solution of the problem to the point to which our text[1] carries the problem, the Arabic, as translated by Rosen, adds: 'Or, if you please, you may add the root of four to the moiety of the roots; the sum is seven, which is likewise one of the parts. This is one of the problems which may be solved by addition and subtraction.'

---

[1] Page 108, lines 1–13.

CHAPTER VI

THE Arabic text of Al-Khowarizmi's algebra published by Rosen contains an author's preface which is not found either in the translation published by Libri, or in that by Robert of Chester. As this reveals his conception of the purpose of the algebra, as well as some of the causes which led him to undertake the work, we present it here in the translation by Rosen.[1] Such prefaces in Arabic works usually, just as this one, contained invocations to the Deity and to Mohammed his prophet: in consequence the Christian translators, who were commonly connected with the Church, were wont to leave them out. A summary of the sections in the Arabic text which appear in neither of the Latin translations is also given since the Arabic-English work by Rosen is not widely available, and since these additions show that Al-Khowarizmi had grasped the possibility of the application of the algebra to geometry and trigonometry. This application is frequently neglected to-day by teachers of elementary algebra.

THE AUTHOR'S PREFACE

" In the Name of God, gracious and merciful ! "

" This work was written by Mohammed ben Musa, of Khowarezm. He commences it thus :

" Praised be God for his bounty towards those who deserve it by their virtuous acts : in performing which, as by him prescribed to his adoring creatures, we express our thanks, and render ourselves worthy of the continuance (of his mercy), and preserve ourselves from change : acknowledging his might, bending before his power, and revering his greatness ! He sent Mohammed (on whom may the blessing of God repose !) with the mission of a prophet, long after any messenger from above had appeared, when justice had fallen into neglect, and when the true way of life was sought for in vain. Through him he cured of blindness, and saved through him from perdition, and increased through him what before was small, and collected through him what before was scattered. Praised be God our Lord ! and may his glory increase, and may all his names be hallowed — besides whom there is no God ; and may his benediction rest on Mohammed the prophet and on his descendants !

[1] Rosen, *The Algebra of Mohammed ben Musa*, pp. 1–4.

"The learned in times which have passed away, and among nations which have ceased to exist, were constantly employed in writing books on the several departments of science and on the various branches of knowledge, bearing in mind those that were to come after them, and hoping for a reward proportionate to their ability, and trusting that their endeavors would meet with acknowledgement, attention, and remembrance — content as they were even with a small degree of praise ; small, if compared with the pains which they had undergone, and the difficulties which they had encountered in revealing the secrets and obscurities of science.

"Some applied themselves to obtain information which was not known before them, and left it to posterity ; others commented upon the difficulties in the works left by their predecessors, and defined the best method (of study), or rendered the access (to science) easier or placed it more within reach ; others again discovered mistakes in preceding works, and arranged that which was confused, or adjusted what was irregular, and corrected the faults of their fellow-laborers, without arrogance towards them, or taking pride in what they did themselves.

"That fondness for science, by which God has distinguished the Imam al Mamun, the Commander of the Faithful (besides the caliphat which He has vouchsafed unto him by lawful succession, in the robe of which He has invested him, and with the honours of which He has adorned him), that affability and condescension which he shows to the learned, that promptitude with which he protects and supports them in the elucidation of obscurities and in the removal of difficulties, — has encouraged[1] me to compose a short work on Calculating by (the rules of) Completion and Reduction, confining it to what is easiest and most useful in arithmetic, such as men constantly require in cases of inheritance, legacies, partition, law-suits, and trade, and in all their dealings with one another, or where the measuring of lands, the digging of canals, geometrical computation, and other objects of various sorts and kinds are concerned — relying on the goodness of my intention therein, and hoping that the learned will reward it, by obtaining (for me) through their prayers the excellence of the Divine mercy : in requital of which, may the choicest blessings and the abundant bounty of God be theirs !   My confidence rests with God, in this as in everything, and in Him I put my trust.   He is the Lord of the Sublime Throne. May His blessing descend upon all the prophets and heavenly messengers ! "

The Arabic version differs from the Latin translations which have come down to us, in giving an extended discussion of inheritance problems and also in discussing geometrical measurements. In the English translation by Rosen the inheritance problems, involving largely legal questions rather than algebraical ones, occupy 79 pages as opposed to 70 for the algebra proper.   The mensuration problems take some sixteen pages of text.   The formulas are

---

[1] Several writers have asserted that the work of Al-Khowarizmi was written at the request of the caliph.   The text shows that this is not Al-Khowarizmi's statement of the case.

See Woepcke, *Extrait du Fakhri*, p. 2 ; A. Marre, *Le Messâhat de Mohammed ben Moussa al Khàrezmì, extrait de son algèbre*, in *Annali di matematica*, Vol. VII, first series, Rome, 1865, pp. 269–280.

given for the area of a square and triangle.   Three formulas are
given for the circumference of a circle and the writer evidently
recognizes them all as approximations.   The formulas are:

(1) $$c = 3\tfrac{1}{7}d,$$
(2) $$c = \sqrt{10\,d^2}$$
(3) $$c = \frac{62832\,d}{20000}$$

The area of a circle is given as $A = d^2 - \tfrac{1}{7}d^2 - \tfrac{1}{2}$ of $\tfrac{1}{7}d^2$.   Other
simple areas and volumes are discussed.   Application of the
algebra is found in two problems.   One of these deals with finding
the altitude of a triangle of which the sides are given; the other
with inscribing a square in a given triangle.

As the problems, on finding the altitude of a triangle, being
given the lengths of the sides, and on inscribing in an isosceles
triangle a square, show that Al-Khowarizmi had an appreciation
of the possibilities of the algebra, I present one of the problems,
following Rosen's translation.

"If some one says: 'There is a triangular piece of land, two of its sides having
10 yards each, and the basis 12; what must be the length of one side of a quadrate
situated within such a triangle?' the solution is this.   At first you ascertain the
height of the triangle, by multiplying the moiety of the basis, (which is six) by itself,
and subtracting the product, which is thirty-six, from one of the two short sides
multiplied by itself, which is one-hundred; the remainder is sixty-four; take the
root from this; it is eight.   This is the height of the triangle.   Its area is, therefore,
forty-eight yards: such being the product of the height multiplied by the moiety of
the basis, which is six.   Now we assume that one side of the quadrate inquired for
is thing.   We multiply it by itself; thus it becomes a square, which we keep in mind.
We know that there must remain two triangles on the two sides of the quadrate, and
one above it.   The two triangles on both sides of it are equal to each other: both
having the same height and being rectangular.   You find their area by multiplying
thing by six less half a thing, which gives six things less half a square.   This is the
area of both the triangles on the two sides of the quadrate together.   The area of
the upper triangle will be found by multiplying
eight less thing, which is the height, by half
one thing.   The product is four things less
half a square.   This altogether is equal to
the area of the quadrate plus that of the three
triangles: or, ten things are equal to forty-
eight, which is the area of the great triangle.
One thing from this is four yards and four-
fifths of a yard; and this is the length of any
side of the quadrate.   Here is the figure: "

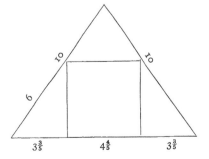

The inheritance problems occupy a large part of the original work; the inclusion of one of these problems here will perhaps not be amiss. Only the first of the problems is given since the following problems are of the same general nature, involving other legal peculiarities.

"A man dies, leaving two sons behind him, and bequeathing one-third of his capital to a stranger. He leaves ten dirhems of property and a claim of ten dirhems upon one of the sons.

"Computation: You call the sum which is taken out of the debt thing. Add this to the capital which is ten dirhems. The sum is ten and thing. Subtract one-third of this, since he has bequeathed one-third of his property, that is, three dirhems and one-third of thing. The remainder is six dirhems (and two-thirds) and two-thirds of thing. Divide this between the two sons. The portion of each of them is three dirhems and one-third plus one-third of thing. This is equal to the thing which was sought for. Reduce it, by removing one-third from thing, on account of the other third of thing. There remain two-thirds of thing, equal to three dirhems and one-third. It is then only required that you complete the thing, by adding to it as much as one-half of the same; accordingly, you add to three and one-third as much as one-half of them: This gives five dirhems, which is the thing that is taken out of the debts."

The legal point involved in the problem given is that a son who owes to the estate of his father an amount greater than the son's portion of the estate, retains, in any event, the whole sum which he owes. Part is regarded as his share of the estate, and the remainder as a gift from the father. The above problem would have given exactly the same numerical results for any debt from five dirhems up; however, if there were a claim of four dirhems against one of the sons, instead of ten, the debtor son would have received in cash $\frac{2}{3}$ of one dirhem, the other son four and $\frac{2}{3}$ dirhems, and the stranger four and $\frac{2}{3}$ dirhems.

In algebraical symbolism, the equation is $\frac{2}{3}(10+x) = 2x$ whence $x = 5$; $10 + x$ is the total estate left, and x is the share of each son.

Plate III.

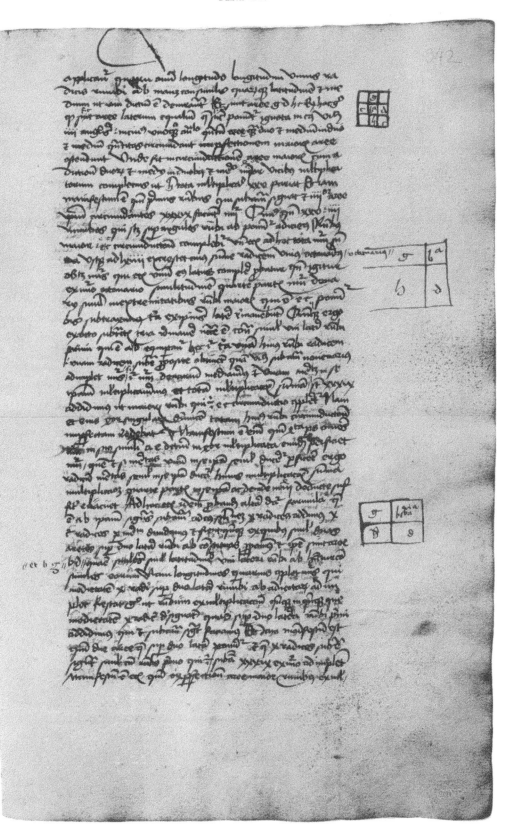

Codex Dresdensis C. 80, Fol. 342 ª

CHAPTER VII

Manuscripts of Robert of Chester's Translation of Al-Khowarizmi's Algebra

I. THE EXTANT MANUSCRIPTS

Steinschneider[1] was the first in recent times to call attention to the translation of Al-Khowarizmi's algebra made by Robert of Chester. He suggested the desirability of publishing this text, referring to the manuscript in Vienna. To this same manuscript Curtze[2] later, and independently of Steinschneider, directed attention and also suggested the desirability of the publication of the work. Wappler,[3] in 1887, found a second copy of the algebra in a manuscript in Dresden, while some time later David Eugene Smith acquired for the Columbia University Library a manuscript from the hand of Johann Scheybl which contains a third transcription of the algebra.

In addition to these manuscript copies of the text, a fragment of the translation was published by Adrien Romain of Louvain in 1599, in a work bearing the title *Commentaire sur l'algèbre de Mahumed ben Musa el Chowarezmi*. Unfortunately but a fragment of this published work had been preserved to modern times, and that precious fragment was doubtless destroyed with other and rarer books and manuscripts in the recent destruction of the University at Louvain. This work of Romain's was mentioned in a work of 1643, published at Louvain, as being found in the Library there. Henri Bosmans, S. J., of Brussels has given a description[4] of the

---

[1] Steinschneider, *Zeitschrift d. deutsehen morgenländ. Gesellschaft*, Vol. XXV (1871), p. 104; *Zeitschrift f. Mathematik*, Vol. XVI (1871), pp. 392–393; *Bibliotheca Mathematica*, third series, Vol. I (1900), pp. 273–274; *Sitzungsbericht d. Akad. d. Wissenschaften in Wien, Phil. hist. Kl.*, Vol. CXLIX (1904), p. 72.

[2] Curtze, *Centralblatt fur Bibliothekswesen*, Vol. XVI (1899), p. 289.

[3] Wappler, *Zur Geschichte der deutschen Algebra im 15. Jahrhundert*, Programm (Zwickau, 1887), pp. 1–2.

[4] Bosmans, *Le fragment du commentaire d'Adrien Romain sur l'algèbre de Mahumed ben Musa el-Chowarezmi*, Annales de la Société scientifique de Bruxelles, Vol. XXX, part II (1906).

work, mentioning the fact noted by Romain that the latter had obtained an excellent manuscript copy of the algebra from his friend Thaddeus Hagec of Prague. I am indeed fortunate, through the courtesy of Professor Bosmans and of Professor B. Lefebvre of Louvain, to be able to include this fragment in my collation. Another fragment in Ms. was found after the textual notes were in type; this brief portion from Codex Dresdensis C. 80^m, of the fifteenth or sixteenth century, is included later in this chapter.

Yet another fragment of the algebra was published by Wappler,[1] who did not, however, ascribe the passage to Robert of Chester. The failure to connect it directly with the algebra in the same manuscript elsewhere mentioned by Wappler was due in part to the fact that the section in question, which is found p. 120, l. 21 to p. 124, l. 16 of our text, is not in its proper place in the Dresden manuscript. The same paragraphs are found also in a manuscript of the University Library at Leipzig, Codex Lips. 1470. I am indebted for information concerning this manuscript to the courtesy of Director Boysen and Dr. Helssig, of the University Library in Leipzig. It appears that this manuscript is almost entirely from the hand of Magister Virgilius Wellendorfer of Leipzig, written during his student days, between the years 1481 and 1487. The passage referred to was in all probability copied from the Dresden Codex C. 80. Apparently Wellendorfer had the intention of copying the algebra entire, for on folio 478^a we find the title, *Textus algabre edidit Mahume Mosi filius*. To the title he added a brief note, Sed utitur aliis nominibus . . . substantia et dragma, radicis . . . algorithmi. These words clearly indicate some acquaintance with our text, but the text itself is not found in the manuscript.

The Scheybl Ms. (C.), written in 1550, was evidently intended for publication. In printing Robert of Chester's text I have thought it best to follow the Ms. which was prepared by Scheybl. Although it contains some errors, and slight additions by Scheybl, these are quite easily distinguished. The advantage of following Scheybl's careful revision seemed obvious; particularly the chapter divisions and sub-titles, many of which he supplied,

---

[1] Wappler, *Zur Geschichte der Mathematik, Zeitschrift f. Math. und Physik*, Vol. XLV (1900), *Hist. lit. Abth.*, pp. 55–56.

make the text easier to follow. That the reader may have before him all the evidence in regard to the text, the readings of the other Mss. are recorded in the critical notes. These may seem needlessly full on account of the inclusion of many apparently unimportant variants; yet in any attempt to determine the parentage of the Mss. such evidence frequently possesses a significance which is not at first sight apparent.

The Dresden Codex C. 80^m, of which I have photographic copies through the courtesy of the director of the Dresden library, contains a fragment of Robert of Chester's translation of the algebra of Al-Khowarizmi. The passage is introductory to a work on arithmetic; the latter contains portions of the algorism by Sacrobosco, and of the algorism in verse by Alexander de Villa Dei, and also parts of the commentary on the former by Petrus de Dacia (1291), together with other comment and further exposition. The " Rules corresponding to the rules of algebra," which we have reproduced on page 126 from the Vienna Ms. are also included with other algebraic material in this mathematical Ms. Here these rules [1] follow closely the Vienna text, whereas similar rules [2] in Codex Dresden C. 80, fol. 351ᵃ, do not.

The fragment from Codex Dresden C. 80^m follows:

" Incipit liber restauracionis numeri quem edidit machumed filius moysi algauriszmi quare dixit machumued.

"Laus deo creatori qui homini contulit scientiam inveniendi vim numerorum. Considerans enim omne id quo indigent homines ex numero, inveni id totum esse numerum. Et nil aliud esse numerum nisi quod ex vnitatibus componitur. Vnitas ergo est qua vnaquaque res dicitur vna et vnitas in omni numero reperitur. Inveni autem omnem numerum essentialiter ita dispositum vt omnis numerus vnitatem excedat vsque ad 10. Decenus quoque numerus ad modum vnitatis disponitur, vnde et duplicatur et triplicatur quemadmodum factum est ex vnitate. Fiuntque ex eius duplicatione 20, ex eius triplicatione 30. Et sic multiplicando decenum numerum ad centenum peruenitur. Proinde centenus numerus duplicatur et triplicatur, etc. ad modum numeri deceni. Et sic centenum numerum duplicando et triplicando, etc. Millenus excrescat numerus. Ad hunc ergo modum millenus numerus ad modos numerorum, vsque ad infinitam numeri investigationem conuertitur."

In this passage the sentence, " Vnitas ergo est qua vnaquaque res dicitur vna," is interjected from Sacrobosco's algorism; it is a translation of Euclid's definition of a unit. The work which im-

---

[1] Published by Wappler, *Programm* (Zwickau, 1887), note 1, page 14.

[2] Published by Wappler, *loc. cit.*, pp. 13–14.

mediately follows concerns arithmetic proper, and the remainder of the material in the Ms. is also mathematical.

The Vienna Ms. (V) is assigned in the catalogue of Mss. of the Vienna library to the fourteenth century, and the character of the writing agrees with this dating. In the fifteenth century Peurbach and Regiomontanus, diligent students of Arabic mathematics, were connected with the University of Vienna, which was then a mathematical centre. This copy itself may have been acquired by Regiomontanus for the library at Vienna; other Mss. from the hand of Regiomontanus are preserved in Vienna.

The Dresden Codex C. 80 (D) was written in the latter part of the fifteenth century. Wappler[1] states that Johann Widmann of Eger, whose activity at the University of Leipzig falls at the end of the fifteenth century, transcribed certain portions of the Dresden Codex, but the writer of our text is not known. Adam Riese, in the beginning of the sixteenth century, also used this Ms. The algebra of our text begins on folio 340^a and terminates in the middle of folio 348^b; the section relating to commercial problems is not found in its proper place, but appears in the Ms. on folio 301^a. For the collation of this latter section I have used the printed text, mentioned above, by Wappler.

II. THE VIENNA MANUSCRIPT (V)

Codex Vindobonensis 4770 (Rec. 3246) XIV. 339.8° (V).

Fol. 1^a–12^b, *Liber restaurationis et oppositionis numeri;* our text.

Fol. 13^a–40^a, M. Jordanus Nemorarius, *De numeris datis,* incorrectly designated in the catalogue as the *Tractatus arithmeticus* by the same author. Many manuscript copies are extant, and the text was published by Treutlein in Vol. II, *Abhandlungen zur Geschichte der Mathematik* (Leipzig, 1879), pp. 135–166, but from an incomplete manuscript of the work. The necessary additions are given by Curtze, *Commentar zu dem " Tractatus de Numeris Datis" des Jordanus Nemorarius, Zeitschrift für Mathematik und Physik, Hist.-lit.-Abth.,* Vol. XXXVI, pp. 1–23, 41–63, 81–95, and 121–138.

Fol. 40^b–44^b, blank.

[1] Wappler, *loc. cit., Programm,* p. 9.

Fol. 45ᵃ–50ᵇ, *Tractatus geometricus cum figuris*. Begins, "Punctum est cuius pars non est . . ." and ends, "equaliter distantes fuerint constituti."

Fol. 51ᵃ–173ᵇ, blank.

Fol. 174ᵃ–324ᵇ, *Carmen quadripartitum de matematica cum commentario subnexo*. This is the *Quadripartitum numerorum* by Johannis de Muris, of which many manuscript copies are extant. The work as a whole has never been published. Two chapters of the second book which relate to practical arithmetic have been published by Nagl, *Abhandlungen zur Geschichte der Mathematik*, Vol. V (Leipzig, 188), pp. 135–146. I have made a study of the third book and I have given selections from the metrical portion of the work, as well as a few passages from the third book and from other parts of the work, *Bibliotheca Mathematica*, third series, Vol. XIII (1913), pp. 99–114.

Fol. 325ᵃ–327ᵇ, blank.

Fol. 328ᵃ–337ᵃ, *Tractatus de ponderibus* (in fine mutilus). This treatise begins, "Marcha est limitata ponderis . . ." and ends, "ad tertium altare et deinceps." This is also the work of Jordanus Nemorarius, and has not been edited.

Fol. 337ᵇ–338ᵇ, blank.

Fol. 339ᵃ, *Notabile de algorismo proportionum*.

III. THE DRESDEN MANUSCRIPT (D)

Codex Dresdensis C. 80. Fifteenth century. 416 pages, numbered 1–225, 227–417. Paper. By various hands. Folio, bound in parchment (D).

Page 1, Vnum dat finger brucke duo. . . . etc. Evidently on finger-reckoning.

Page 1ᵇ–5ᵇ. Large portions of the algorism of Sacrobosco.

Page 6ᵃ. *Divisio numeri*.

Page 6ᵇ–9ᵇ, *Schachirica mercatorum computatio*, etc. Various fragments.

Pages 10, 20–23, 84–128, 145–153, 173–176, 186–190, 198–200, 220–233, 246–257, 269–279, 327–339, and 381–384 are blank.

Page 11–19, Arithmetic of Johannis de Muris. This is a fragment of the *Arithmetica communis* of de Muris which was printed in 1515 at Vienna, and again, as *Arithmeticae speculativae Libri duo*, at Mainz in 1538.

Page 24–71[b], Arithmetic of Boethius, of which there are numerous editions, with a critical edition by Friedlein (Leipzig, 1867).

Pages 72–83. Excerpts from the arithmetic of Boethius.

Pages 129–134. *Regule de Alegorismo.* The algorism (of integers) published by Boncompagni (*Trattati d'arithmetica*, Rome, 1857, pp. 25–49).

Pages 135–142[b]. *Sequitur de phisicis.* Probably the discussion of fractions in the preceding algorism.

Pages 142[b]–144. *Carmen de ponderibus.*

Begins, Pondera postremis Veterum memorata libellis; ends, Argentum argento liquidis cum mergitur Vndis.

Pages 154–157. *Incipit liber de Sarracenico et de Limitibus et cetera Orelibacio de Abacis. Latino. Arabico. Sarracenico et de Limitibus.*

Begins, Quoniam ad Raciones quasdam presentis libelli.

Pages 157[b]–166. *De diuersitate fraccionum capitulum primum. De additione et duplatione. . . . De radicum extractione capitulum 14.*

Begins, Quoniam in precedentibus frequenter contingit; ends, Et haec de computacione fraccionum sufficiant.

Pages 167–172. *Incipit Canon Magistri Johannis de Muris super tabula tabularum que dicitur proporcionum.*

Canon by John of Meurs, edited by John of Gmunden in 1433.

Pages 177–185[b]. *Jordanus Nemorarius de minuciis libri duo.*

See Eneström, *Das Bruchrechnen des Jordanus Nemorarius, Bibliotheca Mathematica*, third series, Vol. XIV, pp. 41–54.

Pages 191–197[b]. *Quid sit proporcio capitulum primum. De 3[ci] Manerie proporcionis capitulum secundum . . . De diuisione proporcionum cap. 10. Quantitatem aliquam mensurare.*

Pages 201[a]–206[a], *Algorithmus proportionum*, by Nicholas Oresme. Published by Curtze, *Programm* (Thorn, 1868).

Pages 206[b]–219[b], 234–245[b]. (*De proporcionibus.*)

Pages 258–268[b]. *Regule super rithmachiam.* On the ancient game called *Rythmomachia* of which there are three standard treatises published; see Smith, *Rara Arithmetica*, p. 271, and a description of the game by Smith and Eaton, American Mathematical Monthly, Vol. XVIII (1911), pp. 73–80.

Pages 280[a]–285[b], *Algorithmus minuciarum* by Johannes de

Lineriis (fourteenth century). Printed at Padua, 1483, and Venice, 1540.

Pages 286ᵃ–291ᵃ, 292ᵇ–300ᵇ, 303ᵇ–305ᵃ, 306ᵃ–315ᵇ.

Various fragments, including portions of the *Quadripartitum numerorum* by John of Meurs; see Wappler, *Programm*, pp. 7, 31–32.

Pages 291ᵇ–292ᵃ, *Algorithmus de duplici differencia*.

Page 301ᵃ. The passage of our text relating to commercial transactions, in the handwriting of Johann Widmann of Eger.

Pages 301ᵇ–303ᵃ, Mathematical lecture by Gottfried de Wolack, written 1467 or 1468.

Pages 316ᵃ–323ᵇ. *De numeris datis*, by Jordanus Nemorarius.

Pages 331ᵃ–334ᵇ. *Pro regularum Algabre.*

Pages 340ᵃ–348ᵇ. Our text.

Pages 349ᵃ–365ᵇ. Latin algebra, published by Wappler, *Programm* (Zwickau, 1887), pp. 11–30.

Pages 366–367ᵇ. *Cautelae Magistri Campani ex libro de Algebra siue de Cossa et Censu.*

Pages 368–378ᵇ. An algebra in German, somewhat similar to the preceding Latin algebra; see Wappler, *Programm*, and *Abhandl. z. Geschichte d. Math.*, Vol. IX, pp. 539–540, where it is discussed.

Pages 379–380, various.

Pages 385–397ᵇ. *De mensuratione terrarum et corporum*, translated by Gerard of Cremona. Not published. Includes algebraic problems.

Pages 397ᵇ–406. *Liber augmenti et diminutionis;* published by Libri, *Histoire des sciences mathématiques en Italie*, Vol. I, (Paris, 1838), pp. 304–371.

Pages 406ᵇ–407, 409–417. Various.

The above description is based upon the *Katalog der Handschriften der Königl. Öffentlichen Bibliothek zu Dresden* (Leipzig, 1882) by Schnorr von Carolsfeld, and upon the articles by Wappler, as cited above.

IV. THE COLUMBIA UNIVERSITY MANUSCRIPT (C)

Codex Universitatis Columbiae; Columbia University Library Manuscript, X 512, Sch. 2, Q. 308 pages. (C.)

Pages 1–68, *Breuis ac dilucida regularum Algebrae descriptio autore Joanne Scheubelio, in inclyta Tubingensi academia Euclidis professore ordinario.* This is a briefer treatment of the algebra than that published by Scheybl in 1550, as a preface to the first six books of Euclid, *Euclidis Megarensis, Philosophi Mathematici excellentissimi, sex libri priores. . . . Algebrae porro Regulae, propter numerorum exempla, passim propositionibus adiecta, his libris praemissae sunt, eaedemque demonstratae* (Basle, 1550). This algebra was published, separate from the Euclid, in Paris, in 1551. The text of the algebra in the Columbia manuscript has not been published.

Pages 69–70, blank.

Pages 71–122, *Liber Algebrae et Almucabola, continens demonstrationes aequationum regularum Algebrae.* Our text.

Pages 123–157, *Addita quaedam pro declaratione Algebrae*, by Scheybl. This is explanatory of the preceding. It is printed on pages 128 to 156 of this book.

Page 158, blank.

Pages 159–308, *Liber Jordani Nemorarii de datis in quatuor partes digestus.* This is not the complete text of the *De numeris datis*, but contains the statement of the problems and their solutions according to the rules of algebra. In these solutions Scheybl introduces the use of N for number, *co.* for *cosa*, for the first power of the unknown, and a symbol which is very similar to the square root sign for the second power of the unknown, or *substantia*.

V. THE RELATIONS OF THE MANUSCRIPTS

The determination of the relation between the Vienna and Dresden Mss., as well as the relation of the Scheybl Ms. and the two fragments to each of these, is based not only upon a study of particular words and phrases but in large measure upon the omissions made by the various scribes in copying. The main test used is, appropriately enough for these mathematical Mss., an arithmetical one.[1] In connection with the omissions we may observe, as noted by Havet,[2] that a copyist passes easily by error from a given ending or word to a similar ending or word which

---

[1] A. C. Clark, *The Primitive Text of the Gospel and Acts* (Oxford, 1914), pp. i–vii and 1–10.

[2] L. Havet, *Manuel de critique verbale appliquée aux textes latins* (Paris, 1911), pp. 130, 200.

occurs later in the contiguous text; this type of error is the most frequent one made in copying Mss. The "jump from like to like" is particularly prone to occur when the parallel passages recur in similar parts of neighboring lines; in this event the omission approximates the length of a line. Another frequent error in copying is to omit entirely one or more lines or to repeat a whole line. An examination of the lengths of all such passages leads to a fairly definite notion of the length of line of the parent manuscript. It need hardly be stated that any omission is more readily made when the appearance, at least, of sense is preserved after the omission. In our mathematical text the recurrence of similar words and phrases is frequent, and in many places the omission of a line is possible with the preservation of a measure of meaning.

In Mss. which antedate the tenth century, the length of any omission is determined with comparative ease as a certain number of letters, for few abbreviations, and those of a standard type, were used. However, in Mss. of the type of the Dresden and Vienna Mss., with which we are primarily concerned, the length in letters of any omission is by no means a fixed and definite quantity. The abbreviations used by the copyists of the twelfth to the fifteenth century commonly varied, not only from page to page, but even from line to line. On the first page of the Vienna Ms. (Plate I), where the conscious effort would be made, probably, to be uniform in notation, the copyist wrote *t'bus* in line 18, and *tribus* in line 26, while on the following page he wrote *tb9*, and elsewhere he writes 3^{b}9. Another possible form is *trib9* although this does not appear in our Ms. This word then could count either for 3, 4, 5, or 6 letters. Similarly on this same first page (line 11) *duplicatione* is written *duplica^{oe}*, while *triplicatione*, which immediately follows, is written *t'plicatione*. In the Dresden Ms. there are entirely similar variations. Thus in line 5 of fol. 340^{b} (Plate II) 5^{a} appears for *quinta*, in line 7 *q'nte* for *quintae*, *quinario* in line 15, and *qūiq3* and *qūq3* in lines 29 and 32 for *quinque;* in line 37 *substanciam* appears in full and in the next line as *subãm*.

In counting the number of letters in any omission I have assumed the common abbreviations used in the contiguous passages of the Vienna Ms.; for as that is quite certainly the oldest of the Mss. which we are examining, it is in all probability more nearly like the parent Ms. whose existence is established with considerable certainty by our study. On the first page of the Vienna Ms. (Plate I) the line varies in length from 35 to 48 letters, with an average of $40\frac{1}{4}$ letters, while on the first half of fol. 340^{b} of the Dresden Ms. the line varies from 40 to 52 letters with an average of $45\frac{1}{3}$ letters. These facts give an indication of the latitude in variation which we may assume in the line length of the parent Ms.

The omissions of the Vienna Ms. will first be investigated as that is the oldest of our texts. In the text of page 84, lines 7–9, the parent Ms. evidently read:

Sed linea b h similis est linee g d. Nam quoniam linea g l et linea h e in quantitate habentur consimiles quoniam linea g l similis est linee d e.

The Vienna scribe passed from the first *similis est linee* to the second. In the text of the same page, lines 12–13:

Area igitur quam linee *t e, e a* circumdant similis est aree quam linee *m l, l d* circumdant. Area igitur . . .

The scribe dropped from the one *Area igitur* to the other.

In the text given in lines 8–9 of the footnote to lines 1–12, page 88:

. . . similis est aree quam *c t, t l* similis quam *t e* circumdant. Area ergo *n z* similis est aree . . . the scribe of the Vienna Ms. passed from the first *similis est aree* to the second. On page 112, text of lines 1–4:

. . . absque xx rebus. Rem quoque in re multiplicata, et erit substantia. Hec insimul iunge et erunt c et due substantie absque xx rebus. The omission can be conceived of here as of the text between *absque* and *absque*, or xx and xx, or between *rebus* and *rebus*, and similarly in the preceding illustrations. So in line 18, page 116 (footnote): in duo et 4^{a} et erunt xx et 4ta, et multiplica v radices in duo et 4ta et erunt 11 res et 4ta. . . . the text is omitted between any word of the first *in duo et 4^{a} et erunt* and the same word where this phrase recurs.

These five passages are in length, roughly, 45, 43, 38, 54 to 60, and 41 letters, respectively. They indicate a line length of about 40 letters in the parent Ms. One omission which is not from like to like appears to be that of a complete line in the text of page 104, lines 22–24: The length of this omission is 45 letters. On page 110, in line 27, there is an omission by the Vienna scribe of some 36 letters, but as the passage is not at all clear in the Dresden Ms., having *unamquamque* where *antequam unamquamque* should have been written, this may have been a deliberate omission. The omission in the text of page 100, line 19, which is short (about 20 letters) may also have been deliberate as the meaning is not clear. Another short omission is found in the text to page 78, line 12.

As these omissions in the Vienna Ms. correspond to portions of the Arabic text, and as all are found in the Dresden Ms., they establish the fact that the Vienna Ms. is not the ancestor of the Dresden Ms.; similarly the omissions of the Dresden Ms. are found in the Vienna Ms., proving that the Dresden could not be the parent Ms. of the Vienna Ms., entirely apart from the fact that the Vienna is quite definitely older.

In the Dresden Ms. some eleven omissions require examination. Of these eight are instances where the scribe has passed from one word or phrase to a similar word or phrase recurring in the adjacent text. Such omissions are, as we have noted, quite likely to be about the length of the line of the parent Ms., but may be materially shorter or longer. Five of these omissions also suggest a line of about forty letters in the parent Ms. Four of these omissions are rectified by marginal additions made by a second hand, but this does not change their value as indicating the length of line in the parent. Four of these omissions are indicated in the footnote to line 6, page 92. The first from *fuerint* to *fuerint;* the second from *sine re* to *sine re;* the third from *procreant* to *procreant,* and so also the fourth although here a 10 intervenes. The lengths in letters are 37, 43, 16, and 38, respectively. In line 7 on the same page is another short omission, from like to like, in the expression, et eius sextam in dragma et eius sextam; the Dresden Ms. omits the last five words. In the first line on the same page we have the evident omission of a full line, no recurring word appearing.

A much longer omission than any yet mentioned is found on page 94, lines 34–36:

. . . perueniet una substantia abiecta. Et si dixerit 10 sine re in re, dicas 10 in re 10 res procreant, et sine re in re substantiam generat diminutivam. Hoc igitur ad 10 res perueniet abiecta substantia.

The scribe passed from *perueniet* to *perueniet* two lines below. Similar omissions, in length about 16 and 25 letters, are found in the text, page 102, line 4, and page 40, line 19. The omission of *A quibus 21 demptis* after *producentur* 25 (page 110, line 8), is not easily explained, but the omission in the text of page 96, line 17, appears to be that of a complete line of 38 letters.

The single illustration of the repetition of a full line or more in either Ms. occurs in the text on page 114, line 29, where the scribe runs back from the *coequantes* at the end of one sentence (line 29), to the *coequantes* at the end of the preceding sentence; the length is 54 letters. A shorter repetition occurs in the text of page 116, line 6. In the text of page 84, lines 12–13:

Area igitur quam linee *t e*, *e a* circumdant similis est aree quam linee *m l*, *l d* circumdant. Area igitur *t a* similis est aree *m d*. . . . we have noted that the Vienna scribe passed from the one *Area igitur* to the second; the Dresden scribe writes *m* after the first *similis est aree*, passing to the second, but he corrects the error. The length here is 43 letters. Since one Ms. omits and the other starts to omit this line, it was possibly omitted in the body of the text of the parent, and supplied in the margin.

The Vienna manuscript contains on the first page two marginal additions, somewhat in the nature of titles (see Plate I). To these there correspond similar additions in the Dresden manuscript, but while the latter continues with numerous other marginal additions the Vienna manuscript does not follow that practise. Among the particularly noteworthy marginal additions by the first hand in the Dresden manuscript are the geometrical figures to be used in connection with the geometrical demonstrations of the solutions of quadratic equations. The Vienna text contains no such figures. But the Dresden figures are evidently not derived directly from the figures in the Arabic text; they give every evidence of being constructed by the writer of the Dresden manuscript upon the basis of the geometrical explanation given in the text. The lettering does not vary greatly from the Arabic, and two of the figures are left quite incomplete.

The second hand has made in the Dresden text several marginal additions based, in general, upon the Vienna manuscript, and once in agreement with Scheybl's reading as opposed to the Vienna reading. Thus on page 66 of our text the marginal notes to words in lines 17 and 18 *alii centenum* and *alii decem*, and on page 68 to a word in line 2, *alii coniunctis*. Four rather long marginal corrections are made, also apparently based upon the Vienna manuscript, in the text to page 92, line 1, in the text as given in note 6 on the same page, in the text given on page 94, lines 34–36, and on page 96, line 17.

We have found then in each of these two manuscripts seven definite indications of a parent manuscript with a line length of between 36 and 54 letters, not very different from the line length in these Mss. themselves. But aside from the line length of the parent there are other indications that the two Mss. have a common parent. In the text of page 86, line 15, both Mss. read *ex duobus et quarta in se ipsis* where the sense requires *ex unitate et medietate in se ipsis;* this error was evidently in the parent Ms., and possibly in the autograph. Similarly in the text of page 106, lines 11 and 14, both Mss. read plainly 36, and that twice, whereas 49 is the correct numerical result here. The scribe of the Dresden Ms. commonly (5 times) writes *hiis* and the other scribe *his*, but in the text of page 76, line 15, this procedure is reversed; the indication is that *hiis* was the form employed in the parent. The Dresden scribe writes *addicias* and the Vienna scribe *adicias*, except in the text of page 76, line 20 where the latter also writes *addicias;* probably this form was used in the parent Ms. In the text of page 102, line 26, the Vienna Ms. reads 10 *sine re,* and the Dresden Ms. *rem*, while the sense requires 10 *res sine substantia.* Either the translation was incorrect in the parent, or, more probably, the passage was illegible. In the text of page 112, line 22, the Vienna Ms. reads 2000, 500, 50 *et* 4ª, for 2550 *et* 4ª; the Dresden Ms. writes the expression in words. This corresponds to a direct translation from the Arabic, for most early mathematical Mss. in Arabic follow the practise of writing numbers in full numeral words, and not in Hindu-Arabic notation.

The discussion of Scheybl's text is somewhat more complicated than that of the two preceding, for Scheybl follows sometimes the one, sometimes the other, and frequently neither, of the two older texts. As we have indicated above, there is probability that the present Vienna text may have been in the library of the University at Vienna when Scheybl was a student there. Further, we know that the Dresden Codex was used by Johann Widmann of Eger towards the end of the fifteenth century when Widmann was lecturing on algebra in the University of Leipzig; and a little later this same manuscript was used by Adam Riese. Although Scheybl was connected as teacher and student with the University of Tübingen, his first work[1] was published at Leipzig. Scheybl may have been familiar, then, with both of our manuscripts, or with the parent. Scheybl's manuscript was prepared with care for the printer, and the few omissions do not throw any light upon his source.

Another difficulty in connection with the Scheybl text is the fact that he took many liberties with the text. I have noted in many places where Scheybl wrote the word which is given in the Vienna and Dresden Mss., and then deleted to substitute a word of his own. Thus, in the text of page 78, line 4, the words *aequalium* and *scilicet* both, though at first written, were crossed out by Scheybl, and on page 80 similar deleted words in the text of lines 3, 8, and 17 correspond also to words in the Vienna and Dresden Mss. In the text of page 76, line 15, of page 82, line 15, and page 100, line 19, the deleted word follows the Dresden Ms. and not the Vienna, while the reverse is true in the text of page 82, line 8, and page 102, line 12. The agreements of Scheybl's Ms. with the Vienna, as opposed to the Dresden readings, number about 125, whereas, the reverse agreements with the Dresden Ms. number about 70; these include phrases as well as single words. Notably, the two concluding paragraphs of the Vienna Ms. with the date and place of the translation, appear in Scheybl's Ms. and not in the Dresden Ms.

In the text of page 68, line 6, after *colligitur* Scheybl omits a passage of some fifty letters ending in *coniungitur;* while this may be an omission from like to like, yet it may also have been deliberately left out as the statement is a repetition, found in the Arabic, of a passage which precedes (lines 1–2). In the text of page 72, line 6, fuerint ut sunt due uel 3ˢ uel plures seu pauciores fuerint Scheybl passes from one *fuerint* to the second, an omission of some 36 letters. The omission of about ninety letters after *exten-*

---

[1] *De numeris et diversis rationibus seu regulis computationum opusculum* (Leipzig, 1545).

ditur in the text of page 96, line 23, appears to be deliberate, as the meaning of the passage is not clear.   So, also, the omission in the text of page 116, line 13, may have been deliberate, as the multiplication is a simple repetition of work which precedes.

Another difficulty in any exact determination of the genesis of Scheybl's text is the fact that he had access to a copy of the algebra of Al-Khowarizmi in the translation which we have designated as the Libri text.   The evidence of this familiarity is found in the *Addita*, written by Scheybl, which are printed on pages 128–156 of this work, for herein are contained portions of the algebra which were not translated by Robert of Chester, notably the problems involving the square root of two hundred (pages 142–144).

Only a fragment remains of the Romain version, constituting about 24 lines of our text.   In this brief space there are some twelve agreements with the Dresden and Vienna Mss., and variations from the Scheybl text.   However, one agreement with the Scheybl manuscript shows either familiarity with Scheybl's work, or a common source other than the Vienna and Dresden Mss. The title " Liber Algebrae et Almucabola, de quaestionibus arithmeticis et geometricis " appears only in the Romain fragment and in Scheybl's text; and somewhat similarly the word "creatori " after " Laus deo " in line 10, page 66, is common to the Romain version and the other Mss. except the Vienna Ms.

The fragment of our text in Codex Dresden C. 80ᵐ, which we have reproduced above, follows exactly none of the other texts. Line omissions do not occur, but the spellings and transpositions agree sometimes with the Vienna, sometimes with the Dresden readings; in one instance *nil* for *nihil*, the agreement is with the Romain fragment as opposed to the Vienna and Dresden C. 80 readings.   This fragment, then, appears also to be based on a parent of the extant Mss.

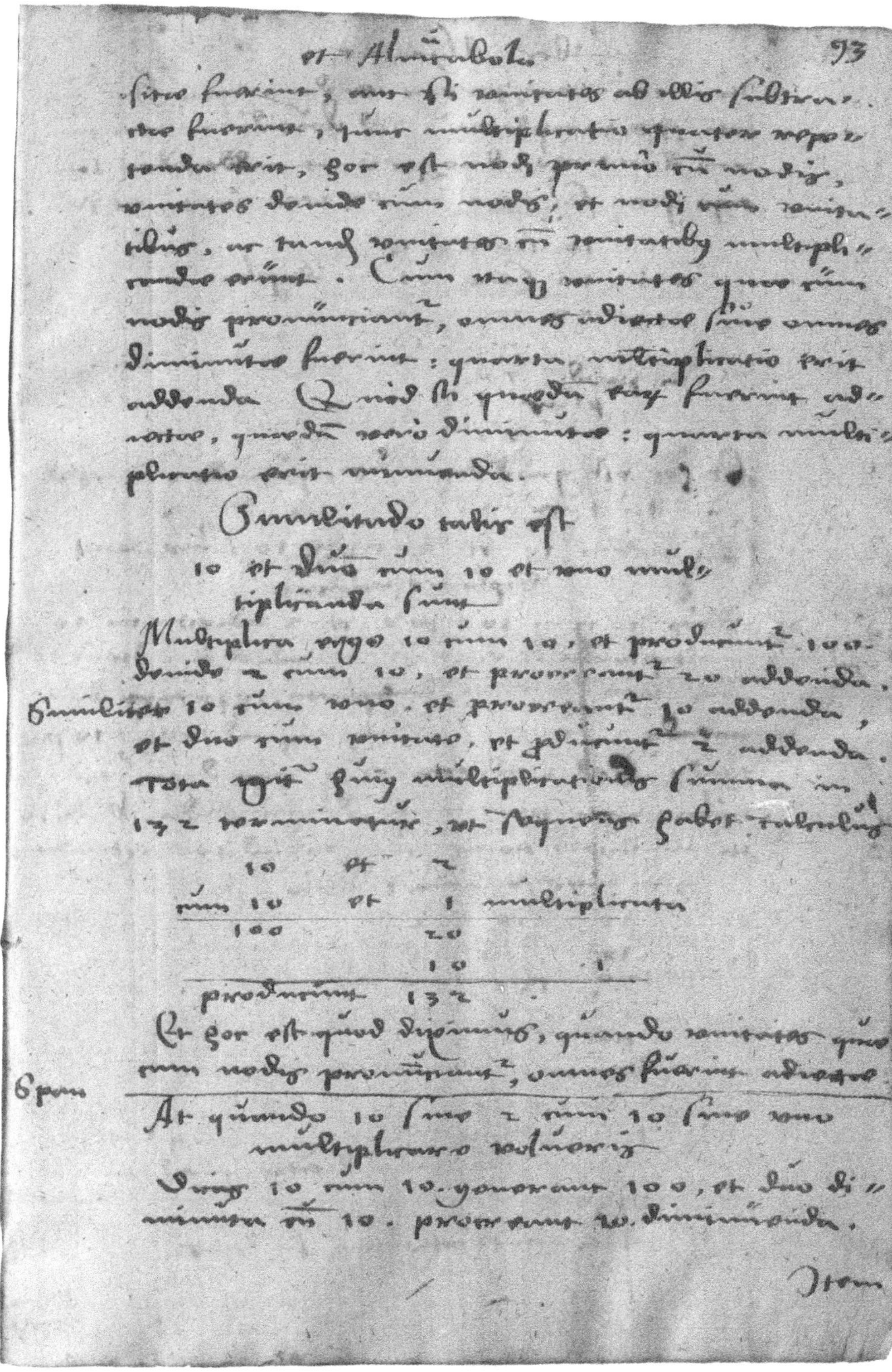

PLATE IV.

et Almucabola

[manuscript text, largely illegible cursive Latin]

Similiter 10 et duo $\overline{\text{cum}}$ 10 et uno multiplicanda sunt

Multiplica ergo 10 $\overline{\text{cum}}$ 10, et producunt 100
donec 2 $\overline{\text{cum}}$ 10, et provenerunt 20 addenda.
Similiter 10 $\overline{\text{cum}}$ uno, et provenerunt 10 addenda,
et duo $\overline{\text{cum}}$ uno, et proveniunt 2 addenda.
Tota igitur huius multiplicationis summa in
132 terminatur, ut sequens habet tabula

	10	et	2	
$\overline{\text{cum}}$	10	et	1	multiplicata
	100		20	
			10	
			1	

producunt 132

Et hoc est quod diximus, quando terminatos $\overline{\text{cum}}$
$\overline{\text{cum}}$ nodis pronunciant, omnes fuerint adiecto

At quando 10 sine 2 $\overline{\text{cum}}$ 10 sine uno
multiplicare volueris

Dicas 10 $\overline{\text{cum}}$ 10. generant 100, et duo diminuta $\overline{\text{cum}}$ 10. provenerunt 20 diminuenda.

Item

ABBREVIATIONS

MANUSCRIPTS

·C = Codex Universitatis Columbiae   R = Fragmentum Romain

D = Codex Dresdensis   V = Codex Vindobonensis

After the note to line 6 of the second page of the Latin text, the notes without any letter indicate the concurrence of V and D.

OTHER ABBREVIATIONS

add. *vel* +	= additum		ras.	= erasura
corr.	= correctum		relict.	= relictum
del.	= deleta		spat.	= spatium
man.	= manus		superscr.	= superscriptum
marg.	' = margine		Tab.	= Tabula
n.	= nota		text.	= textus
om.	= omissum		tit.	= titulum
quaest.	= quaestio		vac.	= vacuum

ϕ = numerus

γ = radix *vel* res

ζ = substantia

$\daleth$ = et

65

LIBER ALGEBRAE ET ALMUCABOLA

CONTINENS DEMONSTRATIONES AEQUATIONUM REGULARUM ALGEBRAE

Ab incerto authore olim arabice conscriptus atque deinde a Roberto Cestrensi, in
5 ciuitate Secobiensi anno 1183, vt fertur, latino sermoni donatus.

LIBER ALGEBRAE ET ALMUCABOLA

Liber Algebrae et Almucabola, de quaestionibus arithmeticis et geometricis.
In nomine dei pii et misericordis incipit liber Restaurationis et Oppositionis
numeri quem aedidit Mahomet, filius Mosi Algaurizin. Dixit Mahomet, Laus deo
10 creatori, qui homini contulit scientiam inueniendi vim numerorum. Considerans
enim omne id quo indigent homines, ex numeris componi, inueni illud totum esse
numerum, et inueni nihil aliud esse numerum, nisi quod ex vnitatibus componitur.
Vnitas ergo in omni numero reperitur. Inueni autem omnem numerum ita dispositum, vt omnis numerus vnitatem excedit vsque ad decem, denarius quoque
15 numerus ad modum vnitatis disponitur, vnde et duplicatur et triplicatur, quemadmodum factum cum vnitate. Fiuntque ex eius duplicatione, 20: et triplicatione, 30. Et sic multiplicando denarium numerum, ad centenarium peruenitur.
Ita centenarius numerus duplicatur et triplicatur, sicut denarius numerus. Et
sic centenarium numerum duplicando et triplicando etc. millenarius excrescit
20 numerus. Ad hunc modum numerum millenarium secundum ordinem numerorum
multiplicando, vsque ad infinitam numeri inuestigationem peruenitur.
Postea inueni numerum restaurationis et oppositionis his tribus modis esse

---

1-6. *om.* VDR.

7. Liber . . . geometricis *om.* VD; + Praefatio R.

8. et [1] *om.* D. instauracionis D.

9. Mahumed filius moysi algaurizim V;
Machumed filius moysi algaurizm D; Mahumed
filius Moysis algaorizim R. Mahomet [2]: Mahumed VR; machumed D.

10. creatori *om.* V.

11. ex numero VDR. componi *om.* VDR.
id *pro* illud V.

12. exlun (?) *sed del. pro* et D. nil R.

13. omnem *om.* D. ita + essentialiter VD;
+ necessario R.

14. vnitatem *corr. ex* vnitas D *man.* 2.
excedat VDR. decenus *pro* denarius *fere ubique*
V; decimus *ubique* D; decenarius *ubique* R.

15. ad *in marg.* D *man.* 2 *pro* ex *del.* triplicatur + et D.

16. est ex *pro* cum VDR. eius *om.* V. et
+ ex VDR.

17. duplicando VDR. centum DR + alii
centenum *in marg.* D *man.* 2; centenum V *et sic
ubique.*

18. Ita: Proinde VD; Post modum R. centenus D. et *om.* V. sicut denarius numerus:
ad modum 10^{ii} (= decimi) V; ad modum numeri
decimi D *et in marg. man.* 2 alii decem; ad modum
decenarii numeri R.

19. decenum *pro* centenarium VD *et corr. in*
centenum D *man.* 2. triplicando multiplicando
duplicando V. etc. *om.* VDR. millenus VD.

20. hunc + ergo VDR. millenus numerus
ad modos VD; millenarius numerus ad modos R.

21. multiplicando *om.* VDR. infiniti V
numeri *om.* V. conuertitur *pro* peruenitur VDR

22. hiis D.

THE BOOK OF ALGEBRA AND ALMUCABOLA[1]

Containing Demonstrations of the Rules of the Equations of Algebra

Written some time ago in Arabic by an unknown author and afterwards, according to tradition in 1183,[2] put into Latin by Robert of Chester in the city of Segovia.

THE BOOK OF ALGEBRA AND ALMUCABOLA

The Book of Algebra and Almucabola, concerning arithmetical and geometrical problems.

In the name of God, tender and compassionate, begins the book of Restoration and Opposition of number put forth by Mohammed Al-Khowarizmi, the son of Moses.[3] Mohammed said, Praise God the creator who has bestowed upon man the power to discover the significance of numbers. Indeed, reflecting that all things which men need require computation, I discovered that all things involve number and I discovered that number is nothing other than that which is composed of units. Unity therefore is implied in every number. Moreover I discovered all numbers to be so arranged that they proceed from unity up to ten. The number ten is treated in the same manner as the unit, and for this reason doubled and tripled just as in the case of unity. Out of its duplication arises 20, and from its triplication 30. And so multiplying the number ten you arrive at one-hundred. Again the number one-hundred is doubled and tripled like the number ten. So by doubling and tripling etc. the number one-hundred grows to one-thousand. In this way multiplying the number one-thousand according to the various denominations of numbers you come even to the investigation of number to infinity.

Furthermore I discovered that the numbers of restoration and opposition

---

[1] *Algebra* and *almucabola* are transliterations of Arabic words meaning 'the restoration,' or 'making whole,' and 'the opposition,' or 'balancing.' The first refers to the transference of negative terms and the second to the combination of like terms which occur in both members or to the combination of like terms in the same member. For a discussion of the terms *algebra* and *almucabola*, see Karpinski, *Algebra*, in *Modern Language Notes*, Vol. XXVII (1913), p. 93. Al-Karkhi included these two operations under *algebra* and the simple equating of the two members as *almucabola*, but Woepcke adds (*Extrait du Fakhri*, p. 64) that this is contrary to the common usage. The title *al-jebr w'almuqabala* is still used in Arabic. The Arabic verb stem *jbr*, from which *algebra* is derived, means 'to restore.' So in Spain and Portugal a surgeon was called an *algebrista*. See also note 3, p. 107.

[2] The date is given in the Spanish Era; 1145 A.D., according to our reckoning.

[3] Mohammed ibn Musa, Al-Khowarizmi. The word *algorism* is derived from his patronymic; the spelling and use in the Latin (see p. 76, line 18), indicate the process of evolution, although the term came into use through Al-Khowarizmi's arithmetic and not his algebra.

1 inuentum, scilicet radicibus, substantiis et numeris. Solus numerus tamen neque radicibus neque substantiis vlla proportione coniunctus est. Earum igitur radix est omnis res ex vnitatibus cum se ipsa multiplicata aut omnis numerus supra vnitatem cum se ipso multiplicatus : aut quod infra vnitatem diminutum 5 cum se ipso multiplicatum reperitur. Substantia verò est omne illud quod ex multiplicatione radicis cum se ipsa colligitur. Ex his igitur tribus modis semper duo sunt sibi inuicem coaequantia, sicut diceres

> Substantiae radices coaequant
> Substantiae numeros coaequant, et
> 10   Radices numeros coaequant.

De substantiis radices coaequantibus. Ca [put] pri [mum].
Substantiae quae radices coaequant sunt, si dicas,
Substantia quinque coaequatur radicibus.
Radix igitur substantiae sunt 5, et 25 ipsam componunt substantiam, quae 15 videlicet suis quinque aequatur radicibus. Et etiam si dicas,
Tertia pars substantiae quatuor aequatur radicibus:
Radix igitur substantiae sunt 12, et 144 ipsam demonstrant substantiam. Et etiam ad similitudinem,
Quinque substantiarum 10 radices coaequantium. Vna igitur substantia dua- 20 bus radicibus aequiparatur, et radix substantiae sunt 2 — et substantiam quaternarius ostendit numerus.
Eodem namque modo, hoc quod ex substantiis excreuerit, aut minus ea fuerit, ad vnam conuertitur substantiam. Et similiter facies cum eo quod cum ipsis ex radicibus fuerit.

25 *De substantiis numeros coaequantibus.* Ca [put] II.
Substantiae verò numeros coaequantes hoc modo proponuntur.
Substantia nouenario coaequatur numero. Nouenarius igitur numerus mensurat substantiam cuius vnam radicem ternarius ostendit numerus. Eodem modo iuxta multitudinem et paucitatem substantiarum ipsae substantiae ad vnius

1–2. id est *pro* scilicet VR ; et in D. Solis numeris VR.   neque *bis supra versum* D *man.* 2, *pro* in *del. bis ;* nec *bis* R.   coniungitur DR *et in marg.* alii coniunctis D *man.* 2 ; coniunctis V.   *In marg.* Quid 𝖅 V ; Radix D.
3. in se *pro* cum se *fere ubique* VDR.   ipsa *om.* V ; ipsam R.
4. ipsa *pro* ipso *fere ubique* D ; ipsum R *passim.*
5–6. vero *om.* D.   ex radicis in se ipsa (ipsa *om.* V ; ipsam R) multiplicacione VDR.   *In marg.* Substantia VD.   *Add. post* colligitur, Solus siquidem (quidem R) numerus neque (nec R) radicibus neque (nec R) substantiis ulla proporcione coniungitur VDR.   ergo *pro* igitur *passim* VD.   semper *om.* VD.
7. semet *pro* sibi.   adinuicem V.   sicuti.
8. coaequant + et.
9–10. et radices numeros coaequant *om.* D.
11. *Titulum om.*

12. Sed et substantie. sunt (+ est D) quasi diceres VD, *et* C *sed del.;* diceres *pro* dicas *fere ubique. In marg.* D 𝖅 φ   coequantur
13. quinque : suis 5.   𝖅   assimilantur
14. 35 V.
15. coequare   *pro* 3 aequare *passim.* quasi diceres.
17. ergo V.
22. ergo *pro* namque.   hoc *om.* V ; his D *et in marg. man.* 2 alii hoc.   eis *pro* ea.
23. conuertas V ; conuertatur D.   facias de.
25. *Titulum om.*
26. In substantiis.   coequantibus.   proponitur D.   *In marg.* 𝖅 assimilantur φ D.
27. equatur V.   numerus *om.* V.
28. cuius scilicet.   Eodem + ergo.
29. modo hoc (hic D) est.   iuxta pluralitatem uel.

are composed of these three kinds: namely, roots, squares[1] and numbers. However number alone is connected neither with roots nor with squares by any ratio. Of these then the root is anything composed of units which can be multiplied by itself, or any number greater than unity multiplied by itself: or that which is found to be diminished below unity when multiplied by itself. The square is that which results from the multiplication of a root by itself.

Of these three forms, then, two may be equal to each other, as for example:

Squares equal to roots,
Squares equal to numbers, and
Roots equal to numbers.[2]

CHAPTER I

Concerning squares equal to roots [3]

The following is an example of squares equal to roots: a square is equal to 5 roots. The root of the square then is 5, and 25 forms its square which, of course, equals five of its roots.[4]

Another example: the third part of a square equals four roots. Then the root of the square is 12 and 144 designates its square.[5] And similarly, five squares equal 10 roots. Therefore one square equals two roots and the root of the square is 2. Four represents the square.[6]

In the same manner then that which involves more than one square, or is less than one, is reduced to one square. Likewise you perform the same operation upon the roots which accompany the squares.

CHAPTER II

Concerning squares equal to numbers [3]

Squares equal to numbers are illustrated in the following manner: a square is equal to nine. Then nine measures the square of which three represents one root.[7]

Whether there are many or few squares they will have to be reduced in the same manner to the form of one square. That is to say, if there

---

[1] Literally 'substances,' being a translation of the Arabic word *mal*, used for the second power of the unknown. Gerard of Cremona used *census*, which has a similar meaning.

[2] These are the three types designated as 'simple' by Omar al-Khayyami, Al-Karkhi, and Leonard of Pisa. They correspond in modern algebraic notation to the following:

$$ax^2 = bx, \; ax^2 = n, \text{ and } bx = n.$$

[3] These and the following chapter headings were doubtless supplied by Scheybl.

[4] $x^2 = 5\,x, \quad x = 5, \quad x^2 = 25.$ [5] $\frac{1}{3}x^2 = 4\,x, \quad x = 12, \quad x^2 = 144.$

[6] $5\,x = 10, \quad x = 2, \quad x^2 = 4.$ [7] $x^2 = 9, \quad x = 3.$

1 substantiae similitudinem erunt tractandae.   Hoc est, si substantiae duae vel tres vel quatuor, siue etiam plures fuerint, earum cum suis radicibus coaequatio, sicut vnius cum sua radice, quaerenda est.   Si verò minus vna fuerit, hoc est, si tertia vel quarta vel quinta pars substantiae vel radicis proposita fuerit, eodem modo ea

5 tractetur, vt si dicas,

Quinque substantiae 80 coaequantur.   Vna igitur substantia quintae parti numeri 80 coaequatur, quam videlicet 16 componunt.   Et etiam si dicas,

Medietas substantiae 18 coaequatur.   Haec igitur substantia 36 coaequatur. Hoc modo omnes substantiae, quotquot in vnum coniunctae, seu quae ab aliis

10 diminutae fuerint, ad vnam conuertentur substantiam.   Hoc idem cum numeris etiam, qui cum substantiis fuerint, agendum est.

De radicibus numeros coaequantibus.   Ca [put] III.

Radices quae numeros coaequant sunt, si dicas,

Radix ternario aequatur numero.   Huius igitur radicis substantiam nouenarius

15 habet numerus.   Et etiam si dicas,

Quatuor radices vigeno coaequantur numero.   Vna igitur radix substantiae huius quinario coaequatur numero.   Et etiam si dicas,

Media radix denario coaequatur numero.   Tota igitur radix vigeno aequatur numero, cuius videlicet substantiam 400 demonstrant.

20   Radices igitur et substantiae et numeri solum, quemadmodum diximus, distinguuntur.   Vnde et ex his tribus modis quos iam praemisimus, tria oriuntur genera tripliciter distincta; vt

Substantia et radices numeros coaequant

Substa[n]tia et numeri radices coaequant, et

25   Radices et numeri substantiam coaequant.

De substantiis et radicibus numeros coaequantibus.   Ca [put] IIII.

Substantiae verò et radices quae numeros coaequant, sunt, si dicas,

Substantia et 10 radices 39 coaequantur drachmis.   Huius igitur artis inuestigatio talis est: dic, quae est substantia, cui si similitudinem decem suarum radicum

30 adiunxeris, ad 39 tota haec collectio protendatur.   Modus hanc artem inueniendi est, vt radices iam pronunciatas per medium diuidas, sed radices in hac interrogatione sunt 10, accipe igitur 5, et iis cum se ipsis multiplicatis producuntur 25;

1. id est *pro* Hoc est V *passim.*   si *om.* D.

2. seu quatuor seu.   radicibus + queratur. equacio V.

4. vel[1] *om.* D. hoc modo tractetur. Sicut diceres.

7. numeri *om.*

9. substantias.   coniunctae + fuerint.

10. conuertas.   quod et de eo quod cum ipsis ex numeris fuerit *pro* Hoc idem . . . fuerint.

11. est *om.* D.   12. *Titulum om.*

13. Sed et radices.   sicut diceres.

14. coequatur *pro* equatur.

16. equantur *pro* coaequantur.   *In marg.*

ꝛ̃q̃ a̅p̃la̅² φ D.

18. equatur *pro* coaequatur.   coequatur *pro* aequatur.

20. soli.   secundum quod iam diximus.

21. modis *om.*   22. Et *pro* vt.

23. Substantie.

24. Et substantie.   radices et coequant D. et² *om.* D.

25. substantias.

26. *Titulum om.* V; Substantias et radicibus numeri coequantibus. D.

27. numerum coequantes.   sicut.

28. dragmatibus *pro* drachmis; dragma *pro* drachma *ubique.*

29. Dic + ergo.

30. xxxviii D.

32. eas V, ea D *pro* iis.   ipsa *pro* ipsis V. multiplica et fient 25.   facio *siue* sum *pro* produco *ubique.*

are two or three or four squares, or even more, the equation formed by them with their roots is to be reduced to the form of one square with its root. Further if there be less than one square, that is if a third or a fourth or a fifth part of a square or root is proposed, this is treated in the same manner.[1]

For example, five squares equal 80. Therefore one square equals the fifth part of the number 80 which, of course, is 16.[2] Or, to take another example, half of a square equals 18. This square therefore equals 36.[3] In like manner all squares, however many, are reduced to one square, or what is less than one is reduced to one square. The same operation must be performed upon the numbers which accompany the squares.

CHAPTER III

Concerning roots equal to numbers

The following is an example of roots equal to numbers: a root is equal to 3. Therefore nine is the square of this root.[4]

Another example: four roots equal 20. Therefore one root of this square is 5.[5] Still another example: half a root is equal to ten. The whole root therefore equals 20, of which, of course, 400 represents the square.[6]

Therefore roots and squares and pure numbers are, as we have shown, distinguished from one another. Whence also from these three kinds which we have just explained, three distinct types of equations [7] are formed involving three elements, as

A square and roots equal to numbers,
A square and numbers equal to roots, and
Roots and numbers equal to a square.[8]

CHAPTER IV

Concerning squares and roots equal to numbers

The following is an example of squares and roots equal to numbers: a square and 10 roots are equal to 39 units. The question therefore in this type of equation is about as follows: what is the square which combined with ten of its roots will give a sum total of 39? The manner of solving this type of equation is to take one-half of the roots just mentioned. Now the roots in the problem before us are 10. Therefore take 5, which multiplied by itself gives 25, an amount which you add to 39, giving 64.

---

[1] Our modern expression "to complete the square," used in algebra, originally meant to make the coefficient of x^2 equal to unity, i.e. make one whole square.

[2] $5 x^2 = 80$; $x^2 = 16$.　　　[3] $\frac{1}{2} x^2 = 18$; $x^2 = 36$.　　　[4] $x = 3$; $x^2 = 9$.

[5] $4 x = 20$; $x = 5$; $x^2 = 25$.　[6] $\frac{1}{2} x = 10$; $x = 20$; $x^2 = 400$.　[7] 'Types of equations' = genera.

[8] Abu Kamil, Omar Al-Khayyami, Al-Karkhi, and Leonard designate these as 'composite' types. In modern notation: $ax^2 + bx = n$;　$ax^2 + n = bx$;　$ax^2 = bx + n$.

1 quae omnia 39 adiicias, et veniunt 64. Huius igitur radice quadrata accepta, quae est 8, ab ea medietatem radicum 5 subtrahas, et manebunt 3. Ternarius igitur numerus huius substantiae vnam ostendit radicem, quae videlicet substantia nouenario dinoscitur numero. Nouem igitur illam componunt substan-
5 tiam.

Similiter quotquot substantiae propositae fuerint, omnes ad vnam conuertas substantiam. Similiter quicquid cum eis ex numeris siue radicibus fuerit, id omne eo modo quo cum substantiis existi, conuertas. Huius autem conuersionis talis est modus, vt si dicas,

10   Duae substantiae et 10 radices 48 drachmis coaequantur.

Huius artis talis est inuestigatio, vt dicas, Quae sunt duae substantiae inuicem collectae, quibus si similitudo 10 radicum earum adiuncta fuerit, ad 48 tota extendatur collectio. Nunc autem oportet, vt duas substantias ad vnam conuertas. Sed iam manifestum est, quoniam vna substantia medietatem duarum designat,
15 igitur omnem rem in hac quaestione tibi propositam, ad medium conuertas, dicendo: Substantia et 5 radices 24 drachmis coaequantur. Modus huius rei talis est, vt dicas, Quae est substantia, cui si quinque suas radices adiunxeris, ad 24 excrescat. Nunc etiam oportet, vt ad regulam supra datam animum conuertas, et diuidas radices per medium et veniunt 2 et vnius medietas, iis cum se ipsis multiplicatis,
20 producuntur 6 et 1/4, illis 24 adiicias, veniunt 30 et vnius 1/4. Postea huius aggregati radicem quadratam accipias, quam scilicet 5 et vnius medietas componunt, ex qua medietatem radicum, 2 1/2, subtrahas et manebunt 3, quae vnam radicem substantiae exprimunt, quam substantiam nouenarius componit numerus. Et si diceretur,
25   Medietas substantiae et quinque radices 28 coaequantur drachmis.

Huius questionis talis est modus, vt dicas, Quae est substantia, cuius medietati si quinque suas radices adiunxeris, tota summa ad 28 excrescat, ita tamen vt substantia quae prius diminuta fuerat, perfecta compleatur. Igitur huius substantiae medietas cum radicibus secum pronunciatis, duplicanda est; veniunt
30 autem,

1. quas (+ suis D) super 39 adicias (addicias *et sic ubique* D) et fient 64. Huius ergo collectionis. quadrata *om., et sic ubique*. assumpta *pro* accepta.
2. quae est: id est V; est D. id est quinque *pro* 5 *et sic passim*. diminuas *pro* subtrahas *passim*. remanebunt: remanere *pro* manere *ubique*. In *marg.* D, *figurae; vide Tab.* II.
3. scilicet D.
4. noscitur. iam *pro* illam.
6. fuerint + ut sunt due uel (uel *om.* D) 3ᵃ uel plures seu (siue D) pauciores fuerint.
7. quotquot V; quidquid D. seu V. ad id cui (cuius D) substantiam conuertisti *pro* id omne . . . existi.
8. ergo *pro* autem.
9. si *om.*
11. Huius + autem D. est hec *pro* talis est. ut si dicas.
12. 10 rx. + unius V; *x* radicum + unius D. In *marg.* 2 ꝛꝗ + 10ꝛ equantur 24. D.

13. ut ad unam substantiam duas conuertas V.
15. et hoc (hic D) est ut dicas *pro* dicendo.
16. substantiam *pro* substantia. 28 *pro* 24 V. coequant V.
17. sui *pro* suas D.
18. dictam *pro* datam. diuide.
19. erunt *pro* veniunt. eas + ergo *pro* iis. in seipsa D. multiplica.
20. et fient 6 et 1/4 (numeri *pro* 1/4 *bis* D) quas videlicet super 24 adicias fientque (fiantque D). summe *pro* aggregati.
21. accipies V. componit D.
22. qua + videlicet. radicis substantiam *pro* radicem substantiae.
24. diceret.
26. est¹ *om.* D. medietati + cui *supra versum* D *man.* 2.
27. rx *pro* radices V. sumas *et* suas *supra versum* D *man.* 2. 25 *pro* 28 V. quod *pro* vt.
29. fietque *pro* veniunt autem.

Having taken then the square root of this which is 8, subtract from it the half of the roots, 5, leaving 3. The number three therefore represents one root of this square, which itself, of course, is 9. Nine therefore gives that square.[1]

Similarly however many squares are proposed all are to be reduced to one square. Similarly also you may reduce whatever numbers or roots accompany them in the same way in which you have reduced the squares.

The following is an example of this reduction: two squares and ten roots equal 48 units.[2] The question therefore in this type of equation is something like this: what are the two squares which when combined are such that if ten roots of them are added, the sum total equals 48? First of all it is necessary that the two squares be reduced to one. But since one square is the half of two, it is at once evident that you should divide by two all the given terms in this problem. This gives a square and 5 roots equal to 24 units. The meaning of this is about as follows: what is the square which amounts to 24 when you add to it 5 of its roots? At the outset it is necessary, recalling the rule above given, that you take one-half of the roots. This gives two and one-half, which multiplied by itself gives $6\frac{1}{4}$. Add this to 24, giving $30\frac{1}{4}$. Take then of this total the square root, which is, of course, $5\frac{1}{2}$. From this subtract half of the roots, $2\frac{1}{2}$, leaving 3, which expresses one root of the square, which itself is 9.

Another possible example: half a square and five roots are equal to 28 units.[3] The import of this problem is something like this: what is the square which is such that when to its half you add five of its roots the sum total amounts to 28? Now however it is necessary that the square, which here is given as less than a whole square, should be completed.[4] Therefore the half of this square together with the roots which accompany it must be doubled. We have then, a square and 10 roots equal to 56 units. There-

---

[1] $x^2 + 10x = 39$; $\frac{1}{2}$ of 10 is 5, 5^2 is 25, $25 + 39 = 64$. $\sqrt{64} = 8$; $8 - 5 = 3$. $x = 3$; $x^2 = 9$.

For the general type $x^2 + bx = n$, the solution is $x = \sqrt{\left(\frac{b}{2}\right)^2 + n} - \frac{b}{2}$; the negative value of the square root is neglected, as that would give a negative root of the equation.

[2] $2x^2 + 10x = 48$, reducing to $x^2 + 5x = 24$; $\frac{1}{2}$ of 5 is $2\frac{1}{2}$, $(2\frac{1}{2})^2 = 6\frac{1}{4}$, $24 + 6\frac{1}{4} = 30\frac{1}{4}$, $\sqrt{30\frac{1}{4}} - 2\frac{1}{2} = 3$.

The general type $ax^2 + bx = n$ is reduced to the preceding by division, giving $x^2 + \frac{b}{a}x = \frac{n}{a}$, and the solution is, as before, $x = \sqrt{\left(\frac{b}{2a}\right)^2 + \frac{n}{a}} - \frac{b}{2a}$.

[3] $\frac{1}{2}x^2 + 5x = 28$, reducing to $x^2 + 10x = 56$. $x = \sqrt{5^2 + 56} - 5$, or $x = 4$. Note that the value of x^2 is not given here as it usually is.

[4] Attention is called to the force of the expression, "completing the square," as here used with the meaning to make the coefficient of the second power of the unknown quantity equal to unity or making one whole square. See also page 81, footnote 1.

1   Substantia et 10 radices 56 drachmis aequales.

Diuide igitur radices per medium, et veniunt 5, quibus cum seipsis multiplicatis, producuntur 25; illa adde 56, et colliguntur 81; huius collecti radicem quadratam accipias, quam nouenarius componit numerus, atque ex ea medietatem radi-
5 cum 5 subtrahas et manebunt 4, substantiae radix.

Hoc modo cum omnibus substantiis quotquot ipsae fuerint, cum radicibus item et drachmis agendum est.

De substantiis et numeris radices coaequantibus.   Ca [put] V.

Propositio huius rei talis est, vt dicas,
10   Substantia et 21 drachmae 10 radicibus coaequantur.

Ad hoc inuestigandum talis datur regula, vt dicas, Quae est substantia, cui si 21 drachmas adiunxeris, tota summa simul decem ipsius substantiae radices exhibeat. Huius quaestionis solutio hoc modo concipitur, vt radices primum per medium diuidas, et veniunt in hoc casu 5, haec cum seipsis multiplicata, produ-
15 cuntur 25. Ex illis 21 drachmas, quas paulo ante cum substantiis commemorauimus, subtrahas, et manebunt 4, horum radicem quadratam accipias, vt sunt 2, quae ex medietate radicum 5 diminuas, et manebunt 3, vnam radicem huius substantiae constituentia, quam scilicet substantiam nouenarius complet numerus. Quod si libuerit, poteris ipsa 2, quae a medietate radicum iam diminuisti, medietati
20 radicum 5 scilicet addere, et veniunt 7; quae vnam substantiae radicem demonstrant quam substantiam 49 adimplent. Cum igitur aliquod huius capitis exemplum tibi propositum fuerit, ipsius modum cum adiectione, quemadmodum dictum est, inuestiga, quem si cum adiectione non inueneris, procul dubio cum diminutione reperies. Hoc enim caput solum adiectione simul et diminutione indiget
25 quod in aliis capitibus praemissis minime reperies.

Sciendum est etiam, quando radices iuxta hoc caput mediaueris, et medietatem deinde cum seipsa multiplicaueris, si quod ex multiplicatione tollitur vel procreatur, minus fuerit drachmis cum substantia pronunciatis: quaestio tibi proposita nulla erit. At si drachmis aequale fuerit vel procreatur vna radix

1. drachmis *om.*   equiparantes *pro* aequales.

2. erunt 5 et postea eas in seipsis multiplica fientque 25. adde igitur eas super 56 eruntque (que *om.* D) 81 *pro* veniunt 5 . . . colliguntur 81.

3. ergo summe *pro* collecti.

4. et *pro* atque.   5. substantiae radix *om.*

6. Hoc + ergo.   fuerint + et.

7. et numeris secum pronunciatis *pro* item et drachmis.

8. *Titulum om.* V;   radicem *pro* radices D.

9. autem huius artis *pro* huius rei.   talis *om.*

V.   *In marg.* |3 ⊤ φ 𝔄   D.

11. hoc + ergo.

12. similis *pro* simul.   radicibus procreetur.

13. Huius + autem.   hac regula *pro* hoc modo. vt *om.*   Primum ergo radices.

14. diuides D.   et fient.   in hoc casu *om.*

14–16. eas ergo in se multiplica et erunt 25. Ex his (hiis D) ergo 21 minue (diminuas D) que cum substantia iam pretaxauimus *pro* haec cum . . . subtrahas.

16. id est *pro* vt sunt.

18. nouenus *pro* nouenarius.

19. Et si volueris.   poteris *om.*   ipsi medietati et (id est D) 5 *pro* medietati radicum 5.

20. addicias et fient 7;   addicias *corr. in* adicias V.   quae vnam: Hec igitur (ergo D) 7 unam.

21. quam + scilicet.   adimplet D.   aliquid. capitis: capitulum *pro* caput *ubique.*   exemplum *om.*

22. addicione D *et sic passim.*   secundum quod.

23. quam D.   24. Hic D.   enim *om.*

25. tribus *pro* aliis.   preter hunc (hoc *superscr.* D *man.* 2) solum in quibus radices mediantur *pro* praemissis.

26. est *om.*   quoniam quando.   eas *pro* medietatem.

27. deinde *om.*   seipsas V; seipsis D.   colligitur V.   vel procreatur *om. et sic infra.*

28. pronunciatis + fuerit.   quaestio + que.

29. facta fuerat adnullata est, ac si ipsa simul dragmatibus fuerit *pro* proposita . . . vel procreatur.   facta C *sed del.*

fore take one-half of the roots, giving 5, which multiplied by itself produces 25. Add this to 56, making 81. Extract the square root of this total, which gives 9, and from this subtract half of the roots, 5, leaving 4 as the root of the square.

In this manner you should perform the same operation upon all squares, however many of them there are, and also upon the roots and the units.

CHAPTER V

Concerning squares and numbers equal to roots

The following is an illustration of this type: a square and 21 units equal 10 roots.[1] The rule for the investigation of this type of equation is as follows: what is the square which is such that when you add 21 units the sum total equals 10 roots of that square? The solution of this type of problem is obtained in the following manner. You take first one-half of the roots, giving in this instance 5, which multiplied by itself gives 25. From 25 subtract the 21 units to which we have just referred in connection with the squares. This gives 4, of which you extract the square root, which is 2. From the half of the roots, or 5, you take 2 away, and 3 remains, constituting one root of this square which itself is, of course, 9.[2]

If you wish you may add to the half of the roots, namely 5, the same 2 which you have just subtracted from the half of the roots. This give 7, which stands for one root of the square, and 49 completes the square.[3] Therefore when any problem of this type is proposed to you, try the solution of it by addition as we have said. If you do not solve it by addition, without doubt you will find it by subtraction. And indeed this type alone requires both addition and subtraction, and this you do not find at all in the preceding types.[4]

You ought to understand also that when you take the half of the roots in this form of equation and then multiply the half by itself, if that which proceeds or results from the multiplication is less than the units above-mentioned as accompanying the square, you have no equation.[5] If equal

---

[1] $x^2 + 21 = 10\,x$. For this type of equation both solutions are presented since both roots are positive. A negative number would not be accepted as a solution by the Arabs of this time, nor indeed were they fully accepted until the time of Descartes.

[2] For the general type, $x^2 + n = bx$ the solution is $x = \dfrac{b}{2} \pm \sqrt{\left(\dfrac{b}{2}\right)^2 - n}$, and both positive and negative values of the radical give positive solutions of the equation proposed. In this problem we have $x^2 + 21 = 10\,x$; $\frac{1}{2}$ of 10 is 5, 5^2 is 25, $25 - 21 = 4$.

$\sqrt{4} = 2$; $5 - 2 = 3$, one root; $5 + 2 = 7$, the other root.

[3] Another use of the expression, " completing the square."

[4] The Vienna MS. defines, 'in which the roots are halved.'

[5] This corresponds to the condition, $b^2 - 4\,ac < 0$, in the equation $ax^2 + bx + c = 0$; in this event the roots are imaginary.

1 substantiae simul etiam medietas radicum, quae cum substantia sunt, pronunciatur, adiectione simul et diminutione abiectis. Quicquid igitur duarum substantiarum, aut plus, aut minus substantia propositum fuerit, ad vnam conuertas substantiam, sicut in primo capite praediximus.

5 *De radicibus et numeris substantiam coaequantibus.* Ca [put] VI.

In hoc capite sic proponitur,

Tres radices et quatuor ex numeris coaequantur substantiae.

Ad hoc inuestigandum talis datur regula, quod scilicet radices per medium diuidas et venit vnum et alterius medietas, hoc deinde cum seipsis multiplices, et 10 producuntur 2 1/4, his 4 ex numeris adiicias, et veniunt 6 1/4, huius postea radicem quadratam accipias, hoc est 2 1/2. Atque eam tandem medietati radicum, vni scilicet et dimidio, adiicias, et veniunt 4, quae vnam substantiae radicem componunt, quam deinde substantiam numerus 16 adimplet. Quicquid igitur plus siue minus substantia tibi propositum fuerit, ad vnam conuertas substantiam.

15 Ex his igitur modis, de quibus in principio libri mentionem fecimus, sunt tres priores, in quibus radices non mediantur, in tribus vero posterioribus vel residuis mediantur radices prout superius liquet.

DIXIT ALGAURIZIN

Sex sunt modi de quibus, quantum ad numeros pertinet, sufficienter diximus. 20 Nunc verò oportet, vt quod per numeros proposuimus, ex geometrica idem verum esse demonstremus. Nostra igitur prima propositio talis est,

Substantia et 10 radices 39 coaequantur drachmis.

Huius probatio est, vt quadratum cuius latera ignorantur, proponamus. Hoc autem quadratum quod loco substantiae ponimus, et eius radicem scire volu-25 mus atque designare. Sit igitur quadratum *a b* cuius vnumquodque latus vnam ostendit radicem. Iam manifestum est, quoniam, quando aliquod eius latus cum

1. similis erit medietati.   sunt *om.*   pronunciantur V.

2. Quidquid.

3. substantiarum, aut minus substantia seu (siue D) maius; seu *pro* siue *ubique* V.

4. iam diximus.

5. *Titulum om.* V.   *In marg.* $\mathbb{R} \, 7 \, \phi \, \mathfrak{z}$ D.

6. hoc + autem.   primum sic proponuntur.

7. Tres + autem.   ex numero uni.

8. Huius igitur *pro* Ad hoc inuestigandum. quantus *pro* quod scilicet.

9. et fiet una radix et.   igitur *pro* deinde. multiplicata, fientque duo et 4ª. Hec igitur (ergo D) super 4 adicias et erunt 6 et 4ª.

10. 6 1/2 C.   ergo summe *pro* postea.

11. id est *pro* hoc est.   duo et medium *pro* 2 1/2 *et sic saepius.*

11–12. quam super medietatem radicum adicias id est super unum et alterius medietatem fientque (que *om.* D) 4 *pro* Atque . . . 4.

12. quae + scilicet D.

13. 16 ergo substantiam adimplent *pro* quam . . . adimplet.   Quotquot V; Quidquid D.   maius *pro* plus.

14. fuerat D.

15. Hiis (His D) ergo 6 modis de; His *sed del. et* Ex his *in marg.* C.

16. primi *pro* priores.   mediant *bis* D. posterioribus vel *om.*

18. + Nunc vero oportet quod numero proposuimus geometrice idem verum esse probemus *ante* Dixit D *man.* 2.   alguarizim V; alghuarizim D.

19. Sex + autem.   numerum.   *In marg.* De $\overline{\chi}$ prima D.

20. numero *pro* per numeros.   ex *om.*   geometrice.

21. probemus.

22. 4 *pro* 10, 29 *pro* 39 V; xxx et nouem D.

23. rumbus *pro* quadratum *ubique.* Rumbum *in text* D *et* rumbum *in marg. man.* 2.

24. Hic igitur rumbus substantiam quam *pro* Hoc . . . ponimus.   cuius radices V.

25. atque *om.*   designet (designetur D).   et ipse est *pro* Sit igitur.   unumquotque V.   vnam + eius.

26. ostendet.   Et iam.   aliquid.

to the units, it follows that a root of the square will be the same as the half of the roots which accompany the square, without either addition or diminution.[1] Whenever a problem is proposed that involves two squares, or more or less than a single square, reduce to one square just as we have indicated in the first chapter.

CHAPTER VI

Concerning roots and numbers equal to a square

An example of this type is proposed as follows: three roots and the number four are equal to a square.[2] The rule for the investigation of this kind of problem is, you see, that you take half of the roots, giving one and one-half; this you multiply by itself, producing $2\frac{1}{4}$. To $2\frac{1}{4}$ add 4, giving $6\frac{1}{4}$, of which you then take the square root, that is, $2\frac{1}{2}$. To $2\frac{1}{2}$ you now add the half of the roots, or $1\frac{1}{2}$, giving 4, which indicates one root of the square. Then 16 completes the square.[3] Now also whatever is proposed to you either more or less than a square, reduce to one square.

Now of the types of equations which we mentioned in the beginning of this book, the first three are such that the roots are not halved, while in the following or remaining three, the roots are halved, as appears above.

GEOMETRICAL DEMONSTRATIONS

We have said enough, says Al-Khowarizmi, so far as numbers are concerned, about the six types of equations. Now, however, it is necessary that we should demonstrate geometrically the truth of the same problems which we have explained in numbers. Therefore our first proposition is this, that a square and 10 roots equal 39 units.

The proof is that we construct a square of unknown sides, and let this square figure represent the square (second power of the unknown) which together with its root you wish to find. Let the square, then, be $a\,b$, of which any side represents one root.

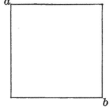

FIG. 1. — This figure appears only in the Columbia manuscript.

---

[1] Condition for equal roots, $b^2 - 4ac = 0$.

[2] $3x + 4 = x^2$; $\frac{1}{2}$ of 3 is $1\frac{1}{2}$, $(1\frac{1}{2})^2 = 2\frac{1}{4}$, $2\frac{1}{4} + 4 = 6\frac{1}{4}$, $\sqrt{6\frac{1}{4}} = 2\frac{1}{2}$, $2\frac{1}{2} + 1\frac{1}{2} = 4$, the root.

[3] The solution of the general type $bx + n = ax^2$, reduced by division to $\frac{b}{a}x + \frac{n}{a} = x^2$, is $x = \sqrt{\left(\frac{b}{2a}\right)^2 + \frac{n}{a}} + \frac{b}{2a}$, and only the positive value of the radical is taken, since the negative value would give a negative root of the proposed equation.

1 numero numerorum multiplicauerimus, tunc hoc quod ex multiplicatione colligitur erit numerus radicum radici ipsius numeri aequalis.   Quoniam ergo decem radices cum substantia pronunciantur, quartam igitur partem numeri decem accipimus, atque vnicuique lateri quadrati aream aequedistantium laterum applicamus quarum
5 longitudinem quidem longitudo quadrati primò descripti, latitudinem verò duo et dimidium, quae sunt numeri 10 quarta pars, demonstrant.   Quatuor igitur areae laterum aequedistantium primo quadrato *a b* applicantur.   Quarum singularum longitudo longitudini vnius radicis quadrati *a b* aequalis erit, latitudo etiam singularum 2 et medium, vt iam dictum est, demonstrat.   Sunt autem hae areae
10 *c d e f*.   Ex hoc igitur quod diximus, sit area laterum inaequalium, quae similiter ponuntur ignota, in cuius videlicet quatuor angulis quorum vniuscuiusque quantitas areae, quam 2 et dimidium cum duobus et dimidio multiplicata perficiunt, imperfectionem maioris seu totius areae ostendunt.   Vnde fit vt circunductionem areae maioris, cum adiectione duorum et dimidii cum duobus et dimidio
15 quatuor vicibus multiplicatorum compleamus, generat autem haec tota multiplicatio 25.   Et iam manifestum est, quoniam primum quadratum, quod substantiam significat, et quatuor areae ipsum quadratum circundantes, 39 perficiunt, quibus quando 25, hoc est quatuor minora quadrata, quae scilicet super quatuor angulos quadrati *a b* ponuntur, adiecerimus, quadratum maius, *G H* vocatum,
20 circunductione complebitur.   Vnde etiam haec tota numeri summa vsque ad 64 excrescet, cuius summae radicem octonarius obtinet numerus, quo etiam vnum eius latus compleri probatur.   Igitur vbi ex numero octonario quartam partem numeri denarii, sicut in extremitatibus quadrati maioris *G H* ponuntur, subtraxerimus bis, 3 ex ipsius latere manebunt.   Quinque ergo ex octo subtractis, 3 manere
25 necesse est; quae simul vni lateri quadrati primi, quod est *a b*, aequiparantur.
   Haec igitur 3 quadrati vnam radicem, hoc est vnam radicem substantiae

1. tollitur D.

2. consimilis *pro* aequalis.   Quando igitur V.

3. substantia primum 10 (xiii D) radices proposuimus 4^{am} partem denarii id est (numeri *pro* id est D) 2 et medium accipiemus (accipiens D) *pro* substantia . . . accipimus.

4. At (ac D) eciam vnicuique 4^{or} laterum rumbi primi aream equalium laterum applicabimus; aequalium *sed del.* C.   quarum + scilicet; scilicet *sed del.* C.

5. quidem *om.*   vnius lateri rumbi primi demonstrent.   Earum (+ earum *in marg.* D *man.* 2) vero latitudinem *pro* quadrati . . . verò.

6. medium *pro* dimidium *ubique.*   numerum obtineant que videlicet 4^{am} partem denarii numeri *pro* quae . . . demonstrant.

7. equalium *pro* aequedistantium.   *a b:* id est *a b.*   omni *pro* singularum.

8. une radicum V.   manet consimilis *pro* aequalis erit.   quarumque latitudinem *pro* latitudo etiam singularum.

9. 2 *om.* D.   demonstrant.   Et hee (*om.* D) sunt.

10. *g d h c et passim.*   sint D, sit D *man.* 2. aree *pro* area D.   equalium *pro* inaequalium.

11. id est in cuius vnoquoque angulo *pro* quorum vniuscuiusque.   quantum D.

12. cum duobus et dimidio *om.* V; in duo et medium quantitas D.   multiplicata *om.* D. circumdant *pro* perficiunt.

13. seu totius *om.*   areae + idem V.   in *pro* vt D

14. in *pro* cum^2.

15. ut *pro* generat autem; perficit *sed del.* C.

16. 25 + pariat.   primus rumbus, qui *et sic passim.*

17. significat: signare *pro* significare *ubique.* quadratum *om.*   perficiunt (faciunt D) + numerum.

18. hoc est quatuor: et 4 V; id est iiii D. minora *om.*   quatuor ^{2} *om.*

19. adiciemus.   id est rumbus (*om.* D) *e c pro G H* vocatum.

20. Vnde etiam + et V; + ad D.

21. vnius octonarius.   qui.

22. Quando igitur *pro* Igitur vbi.   similitudinem 4^e partis denarii numeri simul in *pro* quartam . . . sicut in.

23. qui est *e c pro G H et sic postea.*

25. contra similiter *pro* quae simul.

26. vnam + huius.   id est.

When we multiply any side of this by a number (of numbers) [1] it is evident that that which results from the multiplication will be a number of roots equal to the root of the same number (of the square). Since then ten roots were proposed with the square, we take a fourth part of the number ten and apply to each side of the square an area of equidistant sides, of which the length should be the same as the length of the square first described and the breadth $2\frac{1}{2}$, which is a fourth part of 10. Therefore four areas of equidistant sides are applied to the first square, $a\,b$. Of each of these the length is the length of one root of the square $a\,b$ and also the breadth of each is $2\frac{1}{2}$; as we

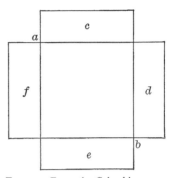

FIG. 2. — From the Columbia manuscript.

have just said. These now are the areas, c, d, e, f. Therefore it follows from what we have said that there will be four areas having sides of unequal length, which also are regarded

FIG. 3. — From the Dresden manuscript.

as unknown. The size of the areas in each of the four corners, which is found by multiplying $2\frac{1}{2}$ by $2\frac{1}{2}$, completes that which is lacking in the larger or whole area. Whence it is that we complete the drawing of the larger area by the addition of the four products, each $2\frac{1}{2}$ by $2\frac{1}{2}$; the whole of this multiplication gives 25.

And now it is evident that the first square figure, which represents the square of the unknown (x^2), and the four surrounding areas $(10\,x)$ make 39. When we add 25 to this, that is, the four smaller squares which indeed are placed at the four angles of the square $a\,b$, the drawing of the larger square, called $G\,H$, is completed. Whence also the sum total of this is 64, of which 8 is the root, and by this is designated one side of the

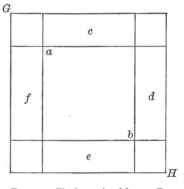

FIG. 4. — Final completed figure. From the Columbia manuscript.

completed figure. Therefore when we subtract from eight twice the fourth part of 10, which is placed at the extremities of the larger square $G\,H$, there will remain but 3. Five being subtracted from 8, 3 necessarily remains, which is equal to one side of the first square $a\,b$.[2]

This three then expresses one root of the square figure, that is, one root of the proposed square of the unknown, and 9 the square itself.

[1] Evidently meaning a pure number.
[2] The proportions of the figures are not correct to scale.

1 propositae: nouenarius deinde numerus ipsam substantiam exprimit. Ergo numerum denarium mediamus, et alteram eius medietatem cum seipsa multiplicamus, deinde totum multiplicationis productum numero 39 adiicimus, vt maioris quadrati *G H* circunductio compleatur. Nam eius quatuor angulorum 5 diminutio totam huius quadrati circunductionem imperfectam reddebat. Manifestum enim est, quod quarta pars omnis numeri cum suo aequali, ac deinde cum quatuor multiplicata, eandem perficiat numerum, quem medietas numeri cum seipsa multiplicata, perficit. Igitur si radicum medietas cum seipsa multiplicetur, huius multiplicationis summa, multiplicationem quartae partis cum seipsa ac 10 deinde cum quatuor multiplicatae, sufficienter euacuet, adaequabit vel delebit.

Ad hoc etiam idem demonstrandum, altera datur formula, quae talis est. Quadrato *a b* substantiam significante aequalitatem decem radicum addimus; has radices deinde per medium diuidamus, venient 5, ex quibus duas areas ad duo latera quadrati *a b* constituamus, hae autem vocentur *a g* et *b d*, et erit vtriusque vtra-15 que latitudo vni lateri quadrati *a b* aequalis; vtramque denique longitudinem numerus quinarius adimplebit. Superest iam, vt ex multiplicatione 5 cum 5, quae medietatem radicum quas ad duo latera quadrati prioris substantiam significantis, addidimus, quadratum faciamus. Vnde iam manifestum est, quòd duae areae, quae supra duo latera ponuntur, et quae 10 radices substantiae significant, 20 simul cum quadrato priori, quod est substantia, 39 ex numero adimpleant. Manifestum etiam, quòd area maioris seu totius quadrati per multiplicationem 5 cum 5 perficiatur. Hoc ergo quadratum perficiatur, atque ad perfectionem eius numerus 25 ad priora 39 adiiciatur: tota igitur haec summa vsque ad 64 excrescet. Nunc summae huius radicem quadratam, quae vnum latus quadrati maioris 25 designat, accipiamus, atque inde aequalitatem eius quod ei addidimus, hoc est 5,

1. proposito obtinent, quam videlicet substantiam nouenarius adimplet numerus (+ si D *man.* 2). Igitur.

2. unam *pro* alteram eius.   in seipsam.   multiplicauimus V; multiplicauerimus D.

3. et *pro* deinde.   totam multiplicationis summam super (sunt D) 39 addidimus; summam 39 C *sed del.*

4. compleretur.   Nam + et.

6. quam *pro* quod *saepius.*   consimili ducta, deinde in 4 V; consimili *a e* ductum in 4ᵒʳ D.

7. eundem perficiet.   quem + si V; + et si D.   medietas + unum D.   numeri *om.*

8. semel duceretur *pro* multiplicata; semel C *sed del.*   perficeret.   si *om.* D.   medietas + semel.   ducatur *pro* multiplicetur; ducere *pro* multiplicare *saepius.*

10. iii D, *sed* 4 *superscr. man.* 2.   adaequabit vel delebit *om.*

11. demonstrandum: comprobandum V; probandum D.   Sit (que est D) *a b* ipsam signans substantiam (+ ad D) cuius similitudinem 10 *pro* quae talis est . . . decem.

12. radices.   10 ergo (igitur D) *pro* has.

13. deinde *om.*   et fient *pro* venient.   quibus + similiter.   super *pro* ad.

14–16. constituas, proponimus et ipse sunt aree *b d* et *b g* (et *b g add. in marg.* D *man.* 2), quare simul (similes *sed del.* D) latitudines (latitudinem D) uni lateri rumbi *a b* habentur consimiles. Earum vero (earum Nam D) longitudines quinarius complet numerus, qui medietatem 10 radicum super duo latera rumbi *a b* adiectarum adimplet.

16. Restat igitur *pro* Superest iam.   vt + rumbum.

17. radicum: 10 radicum designant.   super *pro* ad; sup C *sed del.*   primi D; extremi V. substantiam significantis *om.*

18. addidimus qui et substantiam signet faciamus.   Et iam.   est *om.* V.

20. primo.   adimplent.

21. Manifestum + est.   quam ex perfectione aree *pro* quòd area.   seu totius *om.*   ex multiplicatione.

22. 5² +que (et D) sunt 25 nondum.   perficitur. et *pro* atque.   perfectam *pro* perfectionem.

23. rumbi maioris qui est rumbus *a h* super 39 adiciatur *pro* eius . . . adiiciatur.   Toto ergo D. vsque *om.* V.

24. Huius ergo summe.

25. accipias et ex ea similitudinem eius quod ei addidisti id est 5.

Hence we take half of ten and multiply this by itself. We then add the whole product of the multiplication to 39, that the drawing of the larger square $G H$ may be completed;[1] for the lack of the four corners rendered incomplete the drawing of the whole of this square. Now it is evident that the fourth part of any number multiplied by itself and then multiplied by four gives the same number as half of the number multiplied by itself.[2] Therefore if half of the roots is multiplied by itself, the sum total of this multiplication will wipe out, equal or cancel[3] the multiplication of the fourth part by itself and then by four.

Another method[4] also of demonstrating the same is given in this manner: to the square $a b$ representing the square of the unknown we add ten roots and then take half of these roots, giving 5. From this we construct two areas added to two sides of the square figure $a b$. These again are called $a g$ and $b d$. The breadth of each is equal to the breadth of one side of the square $a b$ and each length is equal to 5. We have now to complete the square by the product of 5 and 5, which, representing the half of the roots, we add to the two sides of the first square figure, which represents the second power of the unknown. Whence it now appears that the two areas which we joined to

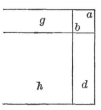

FIG. 5. — Incomplete figure. From the Dresden manuscript.

the two sides, representing ten roots, together with the first square, representing x^2,[5] equals 39. Furthermore it is evident that the area of the larger or whole square is formed by the addition of the product of 5 by 5. This square is completed and for its completion 25 is added to 39.

FIG. 6. — From the Dresden manuscript.

The sum total is 64. Now we take the square root of this, representing one side of the larger square and then we subtract from it the equal of that which we added, namely 5. Three remains,

---

[1] This corresponds to our algebraic process of completing the square. The correspondence of the geometrical procedure to the terminology and the methods employed in algebra makes it highly desirable to present the geometrical and algebraical discussions together to students of elementary mathematics.

[2] $4\left(\dfrac{a}{4}\right)^2 = \left(\dfrac{a}{2}\right)^2.$

[3] The words *adaequabit vel delebit* are doubtless added by Scheybl to explain the force of *euacuet*.

[4] A method slightly different from either of these is given by Abu Kamil and also in the Boncompagni version of Al-Khowarizmi's algebra, ascribed to Gerard of Cremona. This consists in applying to one side of the square a rectangle with its length equal to 10, while the other dimension is the same as that of the square. The two together represent $x^2 + 10x$, or 39. Bisect the side whose length is 10. Now by Euclid II, 6, the square on half the side 10 plus the side of the original square, $(x + 5)^2$, equals the whole rectangle (39) plus the square 25 of half the side 10. The rest of the demonstration is similar to that here given.

[5] I use x^2 for *substantia* here and in the following demonstrations.

₁ subtrahamus, et manebunt 3 ; quae latus quadrati *a b* hoc est vnam radicem substantiae propositae complere probantur. Tria igitur huius substantiae sunt radix ; et substantia nouem.

De substantia et drachmis res coaequantibus.

₅ Substantia et 21 drachmae 10 rebus aequiparantur.

Propositio haec seu quaestio in capite quinto proposita fuit, cuius hic demonstratio docetur. Quadratum igitur *a b*, quod latera habet ignota, substantiam pono, atque ei parallelogrammum rectangulum, cuius vtraque latitudo vni lateri quadrati *a b* aequalis sit, cuiusque longitudo vtraque rerum seu radicum medie₁₀tatem referat, applico. Deinde vero huius rectanguli summam 21 ex numeris constituo, qui numerus cum ipsa substantia propositus est. Haec autem area vel rectangulum *b g* inscribitur, cuius simul latitudo *g d* dinoscitur, longitudo ergo duarum arearum inuicem coniunctarum, in *h d* terminatur. Et iam manifestum est quod haec longitudo denarium obtinet numerum, quoniam omnis area quadrilatera et ₁₅rectorum angulorum, ex multiplicatione sui lateris cum vnitate semel, vnam obtinet radicem : et si cum binario numero, duae eiusdem areae nascuntur radices.

Quoniam igitur primò sic proposuit, vna substantia et 21 drachmae, 10 radicibus aequiparantur, manifestum est, quòd longitudo lateris *h d* in denario terminetur numero, quia latus *h b* vnam substantiae radicem obtinet, latus igitur *h d* ₂₀super punctum *e* per medium diuide. Vnde et linea *e h* lineae *e d* fiet aequalis, atque ducta ex puncto *e* linea perpendiculari *e t* : haec eadem perpendicularis ipsi *h a* lineae aequalis erit. Lineae ergo *e t* portionem, quae ipsaque *d e* linea breuior est, in rectum adiicio *e c*, et fiet linea *t c* aequalis *t g* lineae, vnde quadratum *t l*, quod ex multiplicatione medietatis radicum cum se ipsa multiplicatae, id est ex ₂₅multiplicatione quinarii cum quinario (in hoc casu) colligitur, nobis eueniet. Et iam manifestum etiam est, quoniam area *b g* 21, quae substantiae addidimus, in se obtinet, ex a ea igitur *b g* per lineam *t c*, quae est vnum latus areae *t l* extrahamus atque ex eadem area *b g* aream *b t* minuamus, deinde verò super lineam *e c* quae

1. subtrahas.   quae + simul.   qui *pro* hoc. vnam radicem *om.*   substantia complere probatur. + Hoc igitur (ergo D) unam substantie (substantiam D) radicem adimplet (adimplent D).

3. et *om.* V.   et nouem D.

4. *Titulum om.* V; 21 dragmatibus 4 res D.

5. Substantia + vero.

6. Propositio . . . docetur *om.*

7. habeat.

8. cui V; tibi D; cui C *sed del. pro* atque ei. aream laterum equalium equidistancium V; aream lateris equalium D *pro* parallelogrammum rectangulum.   vtraque *om.*

9. similis; similis *pro* aequalis *saepius.* eiusque V.   vtraque *om.*   ad (*om.* D) quamlibet hanc (*om.* D) quantitatem *pro* rerum . . . referat.

10. Summa igitur huius aree 21 ex numero.

11. qui + simul est.   eciam *pro* autem.   vel rectangulum *om.*

12. *a g.*   inscribi D.   *d in g d om.* C. cognoscitur V.

13. iam *om.* D.   quam sua *pro* quod haec.

14. obtineat.   4ᵗᵃ V; quadrata D.

15. angulorum equalium *pro* rectorum angulorum.   cum *om.* D.   unum *pro* vnitate V. semel *om.* V; semel + deducta D.

16. binario + ducatur.

17. primum.   drachmae *om.*

18. est *om.* D.

19. quoniam *pro* quia.   *a b* V; *h D pro h b.*

20. diuido.   et linee *e d* D.   fiet similis V; similes sunt D.

21–23. atque ducta ex puncto *e* linea . . . in rectum adiicio *e c om.   Add.* Sed et iam manifestum est quam linea *e c* similis sit *h b.* In linea ergo *e t* simile residui *d e* super *e t* adiciam ut et aree circumduccie (circumduccione D) adimpleatur.

24. qui V; est D *pro* quod.   in seipsam V; *om.* D.

25. in hoc casu *om.*   colligitur (colligi D) + et fiunt 25.   nobis eueniet *om.* D.

26. *a g.*   quae + in simul.

27. obtineat.   igitur *om.* D.   *a g pro b g.* linea *pro* per lineam.

28. at V, ac D *pro* atque.   eadem *om.*   at V, ac D *pro* deinde verò.

which proves to be one side of the square $a\,b$, that is, one root of the proposed x^2. Therefore three is the root of this x^2, and x^2 is 9.

CONCERNING A SQUARE AND UNITS EQUAL TO UNKNOWN QUANTITIES

A square and 21 units are equal to ten unknowns. This proposition or problem was proposed in the fifth chapter and here a geometrical demonstration is presented.

Suppose that the square $a\,b$, having unknown sides, represents x^2 and apply to it a rectangular parallelogram of which the breadth is equal to one side of the square $a\,b$ and the length is any quantity you please.[1] Then the numerical value of this rectangle is 21, which number accompanies the same x^2. Moreover this area

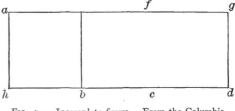

FIG. 7. — Incomplete figure. From the Columbia manuscript, where it twice appears.

FIG. 8.[2] — From the Dresden manuscript.

or rectangle is called $b\,g$, of which one side is $g\,d$ and the length of the two areas together is finally $h\,d$. And it is now evident that this length represents 10, since every quadrilateral having right angles (every square) gives for the product of one of its sides by unity one root, and if multiplied by two gives two roots of its area.

Therefore since the problem was given, x^2 and 21 units equal 10 roots, it is evident that the length of the side $h\,d$ is 10, for the side $h\,b$ designates one root of x^2. Therefore bisect the side $h\,d$ at the point e[3] so that the line $e\,h$ is equal to the line $e\,d$. From the point e draw the perpendicular $e\,t$. This perpendicular equals $h\,a$. Add to the prolongation of the line $e\,t$ a part $e\,c$ equal to the amount by which it is less than $d\,e$ and then $t\,c$ will equal $t\,g$. Whence we arrive at the square $t\,l$ which is the product of half of the roots multiplied by itself, that is the product, in this instance, of 5 and 5. Moreover we know that the area $b\,g$ which we add to x^2 amounts to 21. Therefore we cut across the area $b\,g$ with the line $t\,c$, which is one side of the area $t\,l$, and thus decrease the area $b\,g$ by the amount of the area $b\,t$. Then we form

[1] Scheybl's text is incorrect here.

[2] If the lettering of this figure is made to conform to that of our text the demonstration will be seen to be not materially different; it is based more directly on Euclid II, 5. This proposition, following Heath, *The Thirteen Books of Euclid's Elements*, reads as follows: "If a straight line be cut into equal and unequal segments, the rectangle contained by the unequal segments of the whole together with the square on the straight line between the points of section is equal to the square on the half." This is one of the propositions of Euclid which connect very directly with the geometrical solution of the quadratic equation.

[3] The completed figure (Fig. 9) appears on p. 85. The lettering of Fig. 7 does not correspond to that of the completed figure.

1 est in quo linea *t e c* lineam *a h* in quantitate deuincit, quadratum *e n m c* ponamus. Vnde et iam manifestum, cum linea *t c* aequalis sit linea *l c*, nam ipsae in quadrato *t l* aequales protenduntur, similiter et linea *e c* lineae *m c* aequalis, quia ipsae quadratum *e m* aequali dimensione circundant: aequalium igitur ab aequalibus lineis 5 subtractione facta, linea *t e* lineae *l m* aequalis relinquetur quod est notandum.

Rursus manifestum est, quoniam linea *g d* aequalis est lineae *a h*, cum ipsae in latitudine areae *h g* aequali dimensione tenduntur, sed linea *a h* aequalis est lineae *h b* cum ipsae in vno quadrato appareant. Item quoniam linea *g l* aequalis est lineae *d e*, cum ipsae in vno quadrato reperiantur aequales, sed linea *d e* aequalis 10 est *h e*, cum ipsae decem radices per medium diuidant; linea igitur *d l* residuum lineae *g l*, lineae *e b* ex linea *e h* residuae aequalis erit: atque tandem, cum linea *t e* lineae *l m* ex superiori demonstratione, non sit inaequalis, area quam linea *t e* et *e b* circundant, comprehensae sub *l m* et *d l* lineis areae aequalis erit. Area igitur *t b* aequalis est *m d* areae. Et iam manifestum est, quoniam quadratum 15 *t l* 25 in se continet, cum ergo ex eodem quadrato *t l* areas *d t* et *m d*, quae videlicet duabus areis *g e* et *t b*, 20 et vnum in se continentibus, sunt aequales, subtraxerimus, quadratum *n c* nobis manebit, qui simul numerum, qui est inter 25 et 21, obtinet. Et hic numerus est quaternarius, cuius videlicet radicem duo designant, quae latus *e c* adimplent. Hoc autem latus aequale est lineae *e b*, quoniam *e c* aequale 20 est *d l* lateri, cum ipsa in latitudine areae *d c* aequalia protendantur. Iam manifestum est, quoniam *d l* aequalis est lineae *e b*, quando igitur *e b* quae sunt duo, ex *e h*, quae sunt radicum medietas, quam quinarius ostendit numerus, abstulerimus, linea *b h* ternarium ostendens numerum restabit. Ternarius igitur numerus radicem primae demonstrat substantiae.

25 Quòd si contrà lineam *e c* lineae *e h*, quae medietatem radicum continet, addiderimus, colligentur 7, quae lineam *n h* ostendunt. Et tunc radix substantiae maior

---

1. uincit. *e* et *n m c* faciamus *pro e n m c* ponamus.

2. Sed *pro* Vnde. manifestum est, quam et linea *c t* similis. nam + et.

3. similis est. quoniam et *pro* quia.

4. diuisio *pro* dimensio *ubique*. aequalium . . . facta *om.*

5. linea + igitur V; + ergo D. restat consimilis. quod est notandum *om.*

6. Et iam eciam *pro* Rursus. quoniam + et. similis sit *et sic saepius.* *b h pro a h et sic infra.* quoniam *pro* cum. ipsae + est V.

8. *d e* V; *g d* D *pro h b.* cum ipsae . . . lineae *d e om.* V; Nam quoniam linea *g l* et linea *h e* in quantitate habentur consimiles quoniam linea *g l* similis est linee *d e* D.

9. nam *pro* cum. reperiuntur. aequales *om.* sed *om.* lineaque.

10. Nam eciam et *pro* cum. diuidunt. *d l* residuum lineae *g l*, lineae *om.*

11. *e a pro e b.* residua. erit + linee *d l* ex linea *g l* residue. Sed et *pro* atque tandem, cum.

12. ex superiori demonstratione *om.* iam fuerat consimilis *pro* non sit inaequalis. area quam

t e et *e b* circundant *om.* V; area igitur quam linee *t e, e a* circumdant D.

13. comprehensae sub . . . aequalis erit *om.* V; similis est aree quam linee *m l, l d* circumdant. D.

14. *t a pro t b.* area *pro* quadratum.

15. contineat. area *pro* eodem quadrato. areas + areas D. et *om.*

16. *e t, e g* 21 V; *a t, e g* 20 et unum D. consimiles V; similes D.

17. similitudinem *pro* simul V. numeri.

19. et hoc (hic D). autem *om.* simile est *e a* V; similis est *e a* D *pro* aequale est lineae *e b.* similis *pro* aequale.[2]

20. lateri *om.* Nam et ipse *pro* cum ipsa aequalia *om.* continentur. Et iam.

21. *d* vel (?) *pro d l* D lineae *om.* *e a pro e b.* quando igitur *e b om.* V; Cum *e a* D.

22. radicis. quaternarius D *et* quinarius *in marg. man.* 2.

23. designans *pro* ostendens.

25. Vel *pro* Quòd D. contrà *om.* linea *pro* lineam D. *e n* (*e h* D) super lineam *c h* (*e h* D) addiderimus. quae + solum V; quae + semel D.

26. et (*om.* V) fiunt 7. a[or] *pro* maior D.

the square *e n m c* upon the line *e c*, which is of the length by which the line *t e c* exceeds the line *a h*. Whence, since *t c* equals *l c*, being found in the square *t l*, and similarly since *e c* equals *m c*, these being equal dimensions of the square *e m*, and further equal lines being subtracted from equal lines, it is evident that *t e* is left equal to *l m*. This is to be noted.

Again it is evident that the line *g d* is equal to *a h*, since they represent in breadth equal dimensions of the area *h g*, and the line *a h* equals *h b* as they appear in one square. Also since the line *g l* is equal to *d e*, being found in the same square, and *d e* is equal to *h e*, each being the half of ten roots, therefore the line *d l*, the residuum of the line *g l*, is equal to *e b*, the resi-

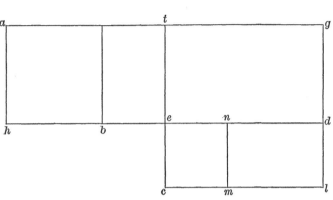

FIG. 9. — Completed figure. From the Columbia manuscript, where it appears twice.

duum of the line *e h*. And so, as the line *t e*, by the above demonstration, is not unequal to the line *l m*, the area which is included by the lines *t e* and *e b* is equal to the area comprehended by the lines *l m* and *d l*. Therefore the area *t b* equals the area *m d*. The square *t l* equals 25. Therefore when we subtract from this same square *t l* the areas *d t* and *m d*, which are of course equal to the two areas *g e* and *t b*, containing 21, it is evident that we have left the square *n c*, which amounts to the difference between 25 and 21. This number is four, of which the root is two, and this gives the line *e c*. Moreover *e c* equals *d l*, since each represents the breadth of the area *d c*. Since *d l* equals *e b*, it is evident that when *e b*, which is two, is taken from *e h*, which is half of the roots, or five, three remains for the line *b h*. Therefore three is the root of the first x^2.

On the contrary if we add the line *e c* to the line *e h*, representing half of the roots, we get 7 which is *n h*. And so the root of the square is greater than (the root of) the

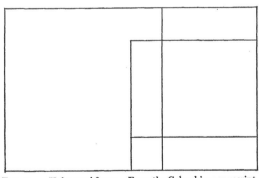

FIG. 10. — Unlettered figure. From the Columbia manuscript.

₁priore substantia erit, quia videlicet super ipsam 21 addidimus. Et similiter ipsa suis 10 radicibus aequalis fiet, quod demonstrare voluimus.

De radicibus et numeris substantiam coaequantibus.

Tres radices et 4 ex numeris coaequant substantiam.

₅ Quadratum igitur cuius latera ignota ponuntur propono, quod sit *a b c d*, atque hoc quadratum tribus radicibus et quatuor ex numeris, vt diximus, aequale constituo. Quoniam autem manifestum est, quòd si vnum latus omnis quadrati semel in vnitatem ducamus vna radix eiusdem quadrati necessario nascatur ex quadrato igitur *a b c d* per lineam *e f* aream *a f* resecemus, atque vnum huius areae ₁₀latus numerum ternarium, id est numerum radicum, significare constituamus. Sit autem hoc latus linea *a e*. Nobis igitur manifestum quoniam area *e c* numero quaternario, quem supra radices adiecimus, adimpletur, igitur super punctum *g* latus *a e* tres radices signans, per lineam in duo media diuidamus, ex quorum vno quadratum, quod est *g k l e* faciamus, quadratum inquam, quod ex multiplicatione ₁₅medietatis radicum in seipsa ductae perficitur, hoc est ex vnitate et medietate cum seipsis multiplicatis producitur.

Deinde lineam *k m* quae sit lineae *e d* aequalis, lineae *g k* adiiciamus, fietque linea *g m* aequalis lineae *g d*, vt quadratum, quod est *g o*, inde nascatur. Et iam manifestum est, quoniam linea *a d* aequalis est lineae *e f*, sed *g d* aequalis est *e n* ₂₀cum ipsae in quadrato *g o* tenduntur aequales, linea igitur *g a* lineae *n f* manebit aequalis. Quia verò *g a* aequalis est lineae *g e* cum ipsae radices medient, id est per medium diuidant, linea insuper *g e* lineae *k l* aequalis, nam ipsae in latitudine areae *e m* continentur aequales: linea igitur *k l* lineae *n f* aequalis erit. Rursus manifestum est, quoniam linea *d g* aequalis est lineae *e n* cum ipsae in quadrato *g o* tendantur ₂₅aequales. Sed linea *g e* aequalis est lineae *e l* cum ipsae quadratum *g l* aequali

1. ista *pro* priore.   erit *om.*   quando *pro* quia. addidimus, fiet ipsa substantia 10 suis consimilis radicibus et hoc est (tamen D) quod explanare voluimus.

2. *Add.*   Sequitur huius alterius partis pro additione figura geometrica C.

3. *Titulum om.* V; De tribus radicibus et 4ᵒʳ ex numero D.

4. Tres + autem.   numero D.

5. ponuntur + substantiam.   sit + rumbus *a d*.   Sed et totum (+ habet D) *pro* atque hoc.

6. numero.   que prediximus.   equalem. constituto D.

7. Et iam *pro* Quoniam autem.

8. simul D.   duxerimus V; duximus D. eius V.   nascetur.   Aream igitur (ergo D) *h d* ex area *a d* resecemus (repetemus D *sed* resecemus *superscr. man.* 2) et unum aliquod (aliud D) *pro* ex quadrato . . . huius areae.

10. signans constituamus.

11. Erit quoque (que D) hoc latus (+ simile D) *z d*.   area *h b*.   4ᵒʳ ex numero que.

12. adimpleat.   *e pro g.*

13. *e h g* V; *h g* D.   per lineam *om.*   quibus *pro* quorum vno.

14. area *e c pro g k l e*.   Eritque hec area que *pro* quadratum inquam, quod.

15. ductae: deducte V; deductam D.   perficitur *om.*   2ᵇᵘˢ et 4ᵃV; duobus et 4ᵗᵃ D *pro* vnitate et medietate.

16. deductis perficitur.

17. *t l pro k m.*   quae sit *om.*   *a h pro e d.* consimilem.   *e t pro g k.*   et ffietque D.

18. *a e pro g m.*   *e l pro g d.*   area *e m pro g o.*

19. *a g pro a d.*   *h z,* sed *h e* (*a e* D).   *h n pro e n.*

20. quoniam *pro* cum.   rumbo *e m.*   *e g pro g a.*   *n c* V; *z n* D.   remansit consimilis *pro* manebit aequalis.

21. Sed et linea *pro* Quia verò.   *e g pro g a. h e pro g e.*   Nam et *pro* cum.   mediant.

22. diuidunt.   quoque *h e* (ergo *h a* sed *a corr. ex e man.* 1 D) similis est linee *t c pro* insuper . . . aequalis.

23. *h l pro e m.*   *n z pro k l.*   *c t pro n f.* Et iam *pro* Rursus.

24. *a h pro d g.*   lineae *om.*   *m n pro e n.* quoniam in latitudine aree *a z* proponuntur *pro* cum ipsae . . . tendantur.

25. Sed + et.   *a e pro g e.*   quoniam *pro* cum.   rumbum *e m* V; rumbos eos D *pro* quadratum *g o.*

first square. Of course when you add 21 to it the sum is equal likewise to ten of its roots which we desired to demonstrate.[1]

CONCERNING ROOTS AND NUMBERS EQUAL TO A SQUARE

Three roots and four are equal to x^2.

I suppose a square, which is $a\ b\ c\ d$, of which the sides are unknown; this square, as we have said, equals three roots and four in number. If one side of any square is multiplied by unity you necessarily obtain one root of the same square. Therefore we cut off from the square $a\ b\ c\ d$ the area $a\ f$ by the line $e\ f$ and one side of this area we take to be three,

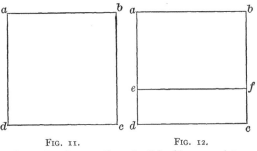

FIG. 11. FIG. 12.

Incomplete figures. From the Columbia manuscript.

constituting the number of the roots; let this side be the line $a\ e$. Now it is clear to us that the area $e\ c$ amounts to the four which is added to the roots. Hence we bisect the side $a\ e$, representing three roots, at the point g. Upon this half construct a square, which is $g\ k\ l\ e$. I say that this square is made by the multiplication of half of the roots by itself, that is, produced by one and one-half multiplied by itself. Then to the line $g\ k$ we add the line $k\ m$, which is equal

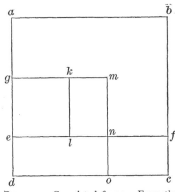

FIG. 13. — Completed figure. From the Columbia manuscript.

to the line $e\ d$. The line $g\ m$ equals the line $g\ d$, thus forming a square which is $g\ o$. Since $a\ d$ is equal to $e\ f$ and $g\ d$ equal to $e\ n$, as they occur in the square $g\ o$, it is evident that $g\ a$ is left equal to $n\ f$. The line $g\ a$ is equal to the line $g\ e$, being half of the roots, that is, bisecting the roots, and it is further true that $g\ e$ equals $k\ l$, for each measures the breadth of the area $e\ m$. Therefore $k\ l$ equals $n\ f$. Again since $d\ g$ equals $e\ n$, being in the square $g\ o$, and $g\ e$ equals $e\ l$, as they measure the same dimension of the square $g\ l$, it is evident that the line $e\ d$ is

[1] This paragraph is not found in the Libri version but appears in the Arabic as published by Rosen. The translation follows the Vienna version. The figure to be used in the geometrical demonstration to obtain by addition the second root of the given quadratic equation appears at the bottom of the preceding page. (Fig. 10.) The Boncompagni version (*loc. cit.* p. 35) varies by letting the middle point fall first within the side of the first square, and secondly without: Cum itaque dividitur per medium linea *b e* ad punctum *z*, cadet ergo inter puncta *g e* aut *b g*: sit hoc prius inter puncta *b g*.

1 dimensione circundant; linea igitur *e d* lineae *l n* aequalis manebit. Quia verò *e d* lineae aequalis est lineae *n o*, cum ipsae in latitudine areae proponantur aequales; et linea igitur *n o* lineae *l n* aequalis erit, atque tandem cum linea *k l* lineae *n f* ex superiore demonstratione non sit inaequalis, area quam lineae *k l* et
5 *l n* circundant, comprehensae sub *n f* et *n o* lineis areae aequalis erit. Area igitur *k n* aequalis est *n c* areae. Et iam manifestum, cum area *e c* quatuor ex numeris, quae supra tres radices addidimus, in se contineat, duae areae *e o* et *k n* vni areae *e c*, quae quatuor ex numeris in se continet, in quantitate aequales fiunt. Manifestum est igitur nobis, quadratum *g o* ex multiplicatione medietatis
10 radicum, id est $1\frac{1}{2}$ ex numeris, cum suo consimili, et adiectione numeri quatuor, cui duae areae *e o* et *k n* aequales sunt, compleri. Est autem totum hoc quadratum numero senario et vni quartae aequale, atque eius radicem duo et medium designant, quae in latere *d g* continentur. Restat igitur nobis ex latere quadrati primi, quod est area *a b c d* quae totam substantiam significat, radicum
15 medietas, quae vnum et medium in se continet, latus etiam *g a*. Cum ergo linea *d g*, quae est latus quadrati *g o*, continentis in se id quod ex multiplicatione medietatis radicum cum seipsa colligitur, cuiusque adiectio sunt quatuor, quae diximus. Et hoc totum sex et vnam quartam in se continet, quarum radicem $2\frac{1}{2}$ super lineam *g a* quae est medietas trium radicum, vnum et medium in se conti-
20 nens, addimus. Tota igitur haec summa, ad quaternarium excrescet numerum, quae est linea *a d*; item radix substantiae atque insuper etiam quadratum *a c*. Tota autem substantia in 16 terminatur. Et hoc est quod exponere voluimus.

Haec igitur geometrica compendiose diximus vt ea quae alioquin oculis mentis difficultate quadam concipiuntur, his geometricè perspectis, ad eadem intelligenda
25 facilior huius disciplinae aditus paretur.

MAHOMET ALGAURIZIN DE ADDITIS ET DIMINUTIS

Et inueni (inquit Mahomet Algoarizin) omnem numerum restaurationis et

1–12. diuisione (+ diuisione D *et* diuisione *superscr.* D *man.* 2) circundant. Et linea *h e* similis est linee *e t*. Nam et ipse rumbum *e c* equali longitudine circumdant. Remanet eciam linea *a h* similis linee *t l*, linea ergo *t l* similis sit (est D) linee *m n*. Sed et linea *c t* linee *n z* iam fuerat consimilis. Area igitur quam *m n*, *n z* circumdant similis est aree (+ quam *c t*, *t l* similis quam *t e* circumdant. Area ergo *n z* similis est aree D; *n z pro n t del.*) *c l*. Et iam manifestum est quoniam area *a z* 4 ex numero que sunt (super D) 3 radices addidimus in se obtineat due igitur aree *a n* et *c l* vni aree *a z* que 4 ex numero in se continet in quantitate fiunt consimiles. Manifestum ergo est nobis quoniam rumbus *e m* qui (numeri D) ex multiplicatione radicum in suo consimili deducte que scilicet duobus (duo habet D) et medium colligitur totum adimpleat (adimplet D) id est duo et $\frac{1}{4}$ que simul rumbum *e c* perficiunt cuius videlicet adiectio sunt 4 ex numero que due aree *a n*, *c l* adimplent. Fietque (Sitque D) hoc totum senario numero et vnius 4^e coequale (coequales D), cuius simul radicem (radicis D) duo et.

13. *a e et sic infra.*
14. *a d.*   quae + et.
15. duo *pro* vnum.   que simul lineam *e g* perficiunt (perfit D).   Cum ergo lineam *a e* que est radix rumbi *e m* que hoc *pro* latus . . . id.
17. radicem D.   in suo consimili + deducte. colligitur + continent.
18. obtinet.   quare radix duo sunt et medium (+ que si V).
19. *e g.*   in se *om.*
20. addiderimus.   igitur *om.*
21. *a g* que est radix substantie que est area *a d.*
22. igitur V; ergo D *pro* autem.   16 + ut subiecta docet descripcio.
23. geometrice.   que oculis mentis quasi quadam difficultate concipiuntur, perspectis his (hiis D) geometrice figuris, ad.
25. faciliorem disciplinem D.   prebeant aditum; aditur paratur C. *Pagina fere tota vac.* V.
26. Mahumed Algorismus V; Mauhumed algaorisim D.   *Titulum om.* C.
27. Mahumed Alguarizmi (algaorismi D).

left equal to *l n*.   Further since *e d* equals *n o*, as they measure the breadth
of the same area, therefore *n o* equals *l n*.   Now as from the above demonstration *k l* is not unequal to *n f*, it follows that the area comprehended by
k l and *l n* equals the area included by the lines *n f* and *n o*.
Therefore the area *k n* equals the area *n c*.   The area *e c*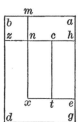
amounts to the number four which we added to the three
roots, and it is evident that the two areas *e o* and *k n* are
equal in quantity to the one area *e c*, containing four in
number.   It is clear to us that the square *g o* consists of the
product of one-half of the roots, *i.e.* $1\frac{1}{2}$, by itself, with the
addition of the two areas *e o* and *k n*.   Thus the sum total
of this square is $6\frac{1}{4}$ and the root of it is given by $2\frac{1}{2}$, which
is contained by the side *d g*.   We have left of the side of

Fig. 14. — From the Vienna manuscript.[1]

the first square, which is the area *a b*, representing x^2, the half of the roots
amounting to $1\frac{1}{2}$ which is the side *g a*.   The line *d g* is a side of the square
g o, containing the product of one-half of the roots by itself with the addition of four, as we have said.   This total amounts to $6\frac{1}{4}$.   We add the root
of this, $2\frac{1}{2}$, to the line *g a* which is one-half of the three roots, amounting
to $1\frac{1}{2}$.   Hence this sum total reaches four, which is the line *a d*.   This is
the root of x^2 and also further of the square *a c*.   The whole of x^2 is finally
16.   This is what we desired to explain.

We have now explained these things concisely by geometry in order that
what is necessary for an understanding of this branch of study might be
made easier.   The things which with some difficulty are conceived of by the
eye of the mind are made clear by geometrical figures.[2]

Positives and Negatives

And I found, says Mohammed Al-Khowarizmi, that all problems of restoration and opposition are included in the six chapters which we have set
forth in the beginning of this book.[3]

---

[1] The writer of the Vienna manuscript took no pains to make the proportions of his figures
correct.   Thus in this figure *a b d g* is intended to represent a square, while *e* is supposed to be
the middle point of the line *h g*.   Further *e c* is to be a square and likewise *e m*, and also the
rectangle *n t* is intended to be equal to the rectangle *z m*.   Moreover the proof in the Vienna
manuscript is not consistent with the lettering of the figure, showing that the copyist did not
succeed in following closely the argument of the text.   Similarly the figures in the Boncompagni
version do not have the correct proportions.

For the proof based directly on Euclid II. 6, see page 133.   The Boncompagni version and
Abu Kamil make explicit reference to the propositions of Euclid.

[2] This paragraph is not found in either the Libri or the Arabic versions; nor in the Boncompagni text.

[3] The evident meaning of this passage is that all problems leading to equations of the first or
second degree can be solved by the methods set forth in the preceding text.

oppositionis in sex capitibus, quae in principio huius libri praemisimus, contineri. Nunc porro, quomodo res vel radices, quando vel solae vel cum illis numeri fuerint, aut quando ex eis numeri extracti, seu cum ipsae ex numeris extractae fuerint, ad inuicem multiplicentur, vel quomodo ad inuicem iungantur, vel ex aliis dimi-
5 nuantur, deinceps dicendum est.

In primis ergo sciendum est, quòd numerus cum numero multiplicari non possit, nisi cum numerus multiplicandus toties sumatur, quoties in numero cum quo ipse multiplicatur, vnitas reperitur. Cum ergo nodi numerorum et cum illis aliquot vnitates propositae fuerint, aut si vnitates ab illis subtractae fuerint, tunc
10 multiplicatio quater repetenda erit; hoc est nodi primò cum nodis, vnitates deinde cum nodis, et nodi cum vnitatibus, ac tandem vnitates cum vnitatibus multiplicandae erunt. Cum itaque vnitates quae cum nodis pronunciantur, omnes adiectae siue omnes diminutae fuerint, quarta multiplicatio erit addenda. Quòd si quaedam earum fuerint adiectae, quaedam verò diminutae; quarta multipli
15 catio erit minuenda.

Similitudo talis est, 10 et duo cum 10 et vno multiplicanda sunt. Multiplica ergo 10 cum 10, et producuntur 100; deinde 2 cum 10, et procreantur 20 addenda. Similiter 10 cum vno, et procreantur 10 addenda, et duo cum vnitate, et producuntur 2 addenda. Tota igitur huius multiplicationis summa in 132 terminatur.
20 Et hoc est quod diximus, quando vnitates quae cum nodis pronunciantur, omnes fuerint adiectae.

At quando 10 sine 2 cum 10 sine vno multiplicare volueris, dicas 10 cum 10 generant 100, et duo diminuta cum 10 procreant 20 diminuenda. Item 10 cum vno procreant 10 diminuta. Hoc autem totum 70 complectitur. Sed duo di
25 minuta cum vno diminuto, duo procreant addenda. Tota ergo haec summa in 72 terminatur. Et hoc est quod diximus, quando omnes diminutae fuerint.

Si autem 10 et 2 cum 10 sine vno multiplicare volueris: dicas 10 cum 10 100,

1. capitulo D.

2. Nunc ergo dicendum (addendum D) est. id est radices quando sole fuerint uel quando cum.

3. uel *pro* aut V.   numeros extraxerint. quando *pro* cum.

4. et qualiter adinuicem iungantur, et qualiter quidam (qui D) ex aliis diminuantur *pro* vel[1] . . . est.

6. est *om.* V.   quam.   in numerum.

7. cum[1] *om.*   duplicandus D.   tociens duplicetur quociens.   in (*om.* D) quem.

8. multiplicatio D.   modi D *saepius; C add.:* seu incerta et obscura cognitio.

10–12. id est nodos in (+ prima D) nodis et unitates in nodis (*om.* D) et iterum unitates in nodis et unitates cum unitatibus. Cum ergo.

13. adiectiuus *pro* adiectus *siue* addendus *et saepius.*   sua *pro* siue D.   diminutiuus *pro* diminutus *siue* diminuendus *et saepius.*

14. alie V; alia D *pro* quaedam[2].   multiplico V; multiplicato D.

16–19. Dicam ergo 10 in 10, 100 procreant, et unum in 10, 10 generat adiectiua et 2 in 10, 20 procreant adiectiua, et duo in 1 duo generant adiectiua.

19. 132 creatur D.   C *add.:* vt sequens habet calculus.   10 et 2

cum   10   et   1 multiplicata
─────────────────
100   20
            10   1
─────────────────
producunt   132

20. qui.

22. 10 sine uno in 10 sine uno (+ in 10 sine uno V).

23–25. et unum diminutum in 10 deductum, 10 procreant diminutiua. Item unum diminutum (diminutiuum D) in 10 deductum, 10 generat diminutiua. Hoc ergo totum 80 (81, *et* 80 C *sed del.*) amplectitur. Sed et unum diminutum in uno diminutiuo unum procreant adiectiuum.

26. C *add.:* vt sequitur,

10   sine   2
cum   10   sine   1 multipli.
─────────────────
100 mi. 20
      mi. 10
─────────────────
Summa prod.   70
        plus   1
─────────────
        71

27. ducas *pro* Dicas D.   fiunt 100 V.

Now further the method is to be explained by which you multiply un-known quantities or roots, either when alone, or when numbers are joined to them or subtracted from them, or when they are subtracted from numbers; also in what manner they are added to, or, in turn, subtracted from, each other.

In the first place you should understand that the only way to multiply a number by a number is to take the number to be multiplied as many times as there are units in the number by which it is to be multiplied.[1] When therefore the nodes[2] of numbers are proposed either with some units or if units are subtracted from them, then the multiplication is fourfold, *i.e.*, first the nodes are multiplied by the nodes, then the units by the nodes and the nodes by the units, and finally the units by the units. When therefore the units which accompany the nodes are both added or both subtracted the fourth product is to be added. But if one is added and the other sub-tracted then the fourth product is to be subtracted.

A problem of this kind is given by the following: 10 and 2 are to be multi-plied by 10 and 1. Hence multiply 10 by 10, giving 100; then 2 by 10, giving 20 to be added; likewise 10 by 1, giving 10 to be added. Two by 1 gives 2 to be added. The sum total of this multiplication is finally 132.[3] And this illustrates what we have said in respect to the type in which the units which accompany the nodes are both to be added.[4]

But when you wish to multiply 10 less 2 by 10 less 1, you say 10 by 10 gives 100; 2 to be subtracted by 10 gives 20 to be subtracted; also 10 by 1 gives 10 to be subtracted. This total, then, amounts to 70. But negative 2 multiplied by negative one gives positive 2. Therefore the sum total is finally 72.[5] This illustrates what we have said when both (bino-mials) involve negatives.

Moreover if you wish to multiply 10 and 2 by 10 less 1, you say 10 by

---

[1] To this definition Al-Khowarizmi refers in his arithmetic (*Trattati*, I, p. 10).

[2] Scheybl adds that this is a word of uncertain and obscure meaning. The Arabic word *'uqud* is connected with the verb meaning "to knot," referring to tying knots on a string to indicate numbers. The Libri and Boncompagni texts use *articuli*, while Rosen translates 'greater numbers.' See also F. Woepcke, in *Journal Asiatique*, Vol. I (6), 1863, p. 276.

[3] The Columbia manuscript continues with the following addition by Scheybl:

The calculation is as follows:

$$\begin{array}{r} 10 + 2 \\ 10 + 1 \\ \hline 100 \quad 20 \\ 10 \\ \hline 132 \end{array}$$

1 [written by mistake for 2]

[4] This is one of the early attempts at a discussion of the multiplication of binominals, including $(x + a)$ times $(x + b)$, $(x - a)$ times $(x - b)$, and $(x - a)$ times $(x + b)$.

[5] Scheybl's text contains by error 81 for 72. The problem as given in the Arabic and Libri versions is $(10 - 1)$ by $(10 - 1)$ with the product 81. This undoubtedly was given by Robert of Chester, and is so recorded in the Vienna and Dresden MSS. Scheybl evidently varied from the text before him, but neglected to make necessary changes in the numerical computation.

1 et 2 adiecta cum 10 multiplicata 20 generant addenda. Item 10 cum vno diminuto multiplicata, 10 procreant diminuenda. Haec autem summa vsque ad 100 et 10 protenditur. Sed 2 adiecta cum vno diminuto multiplicata, 2 procreant diminuenda. Vnde tota multiplicationis summa ad 108 extenditur. Et hoc est
5 quod etiam diximus, quando quaedam earum fuerint adiectae quaedam vero diminutae.

Similiter in fractionibus, si dicas: drachma et eius sexta cum drachma et eius sexta. Dicas, drachma cum drachma drachma[m], et drachmae sexta cum drachma drachmae sextam procreat. Item sexta cum drachma sextam procreat,
10 et sexta cum sexta sextam sextae, id est tricesimam sextam drachmae procreat. Erit autem hoc totum, drachma $\frac{1}{3}$ et $\frac{1}{36}$ drachmae.

Si eodem modo drachma[m] sine sexta cum drachma sine sexta multiplicares, tantum fiet quantum si $\frac{5}{6}$ cum suo aequali multiplicares. Vnde et haec multiplicatio ad 25 partes ex tricesimis sextis partibus vnius drachmae extendetur, id
15 ad $\frac{2}{3}$ et $\frac{1}{6}$ sextae.

Modus autem multiplicationis est, vt drachmam cum drachma multiplices, et producetur drachma; deinde sine sexta cum drachma, sextam procreat diminuendam; item drachmam cum sine sexta, et producetur vna sexta diminuenda. Duae igitur tertiae vnius drachmae supersunt. Et sine sexta cum sine sexta,
20 sextam sextae generat addendam. Tota igitur haec summa ad $\frac{2}{3}$ et sextam sextae extenditur.

---

1. et (*om.* V) unum diminutum in 10 multiplicatum, 10 generat diminutiua, 2 quoque (ergo D) adiectiua in 10 deducta, 20 procreant adiectiua; duo . . . adiectiua² *in marg.* D *man.* 2.

   3. 10 *pro* 2² C.   Hec igitur.

   4. Vnde + et. 180.   C *add.* Haec autem sequenti calculo patent: 10   et   2

10	sine	1
100	plus	20
	mi.	10
Summa pro.		110
	Minus	2

Et *om.*; Est C.   108

   6. diminutiue *et add.* Hoc igitur modo res inuicem multiplicantur. Quando cum ipsis numeri fuerint (+ et quando ipse sine numeris fuerint. Et quando numeri sine ipsis propositi fuerint V). Et si tibi propositum fuerit 10 sine re (+ in 10 deducta quantum constituunt. Dicas 10 in 10, 100 et sine re V; + et expositio rei est re in 10 deducta. Dicas 10 in 10, 100 et sine re *in marg.* D *man.* 2) in 10, 10 generat (generant D) radices diminutiuas. Dicas ergo quod tota hec summa usque ad 100 extenditur 10 rebus abiectis. Et si dixerit 10 et res in 10 deducta quantum procreant. Dicas 10 in 10 centum, et res adiectiua decies deducta 10 res generat adiectiuas. Tota igitur (ergo D) hec summa ad 100 et 10 res extenditur. Si autem dixerit 10 et res in suo consimili quantum multiplicata faciunt. Dicas 10 in 10, 100 procreant (+ et 10 in re 10 res procreant V; + res D). Item 10 in re, 10 res procreant et res

in re substantiam generat adiectiuam. Ergo tota hec summa ad 100 ex numero et 20 res et substantiam extenditur adiectiuam. Si autem sic (si D) proponat, 10 sine re in 10 sine re quantum faciunt. Dicas 10 in 10, 100 et sine re in 10, res procreant 10 (+ diminutiuas et sine re in 10, 10 similiter diminutiuas res procreant V) et sine re in (+ x D) sine re substantiam generat (generant D) adiectiuam. Erit ergo hoc totum 100 et substantia 20 rebus abiectis; *Vide infra pag.* 94.

   7. Eodem modo si dragma (dragmam D) et eius sextam (+ in dragma et eius 6ᵗᵃᵐ V) duxeris, quantum fiet.

   8. et *om.*   dragma in 6ᵗᵃᵐ, sextam dragmatis.

   9. forma *pro* sexta D.

   10. et *pro* cum D. $\frac{1}{36}$ (xxxvi D) dragmatis generat adiectiuam. Erit ergo hoc totum dragma et $\frac{2}{3}$ vel $\frac{1}{3}$ (+ dragma D) et sextam (due D) 6ᵗᵉ.

   12. duxeris.   16. huius *pro* autem. ut dragma.

   17. fietque.   et *pro* deinde.   in dragma + multiplicatum.

   18. et sine 6ᵗᵃ (*om.* D) in dragma, sextam procreat similiter diminutiuam.

   19. vnius *om.*   in sine sexta + deductum.

   21. C *add.*: Sequitur calculus.

         drachma sine sexta
         drach.   sine sex.
         drach.   sine $\frac{1}{6}$
                sine $\frac{1}{6}$

manent $\frac{2}{3}$.   Accedit $\frac{1}{6}$ de $\frac{1}{6}$.

Vnde multiplicationis productum tandem ad $\frac{2}{3}$ plus $\frac{1}{36}$ sese extendat.

10, 100, and positive 2 multiplied by 10 gives positive 20. Also 10 multiplied by negative 1 gives negative 10. This sum, moreover, amounts to 110. But positive 2 multiplied by negative 1 gives negative 2. Whence the sum total of this multiplication equals 108.[1] And this illustrates the type of process when units are to be added and others to be subtracted.[2]

Likewise in the case of fractions, if the problem is a unit and one-sixth (to be multiplied) by a unit and one-sixth. You say, unit by unit gives unit; and one-sixth of a unit by a unit gives one-sixth of a unit. Also one-sixth by a unit gives one-sixth and one-sixth by a sixth gives one-sixth of a sixth, *i.e.* one thirty-sixth. The total will be a unit and $\frac{1}{3}$ and $\frac{1}{36}$ of a unit.

In the same manner, if you multiply a unit less one-sixth by a unit less one-sixth, the product will be the same as $\frac{5}{6}$ multiplied by its equal. Whence this product equals 25 thirty-sixths of one unit, *i.e.* $\frac{2}{3}$ and $\frac{1}{6}$ of one-sixth.

Now the method of this multiplication is that you multiply unit by unit, giving unit; then negative one-sixth by unit, giving negative one-sixth; then you multiply a unit by negative one-sixth, giving one-sixth negative. Therefore two-thirds of one unit remain. And negative one-sixth multiplied by negative one-sixth produces one-sixth of one-sixth positive. The sum total therefore amounts to $\frac{2}{3}$ and one-sixth of one-sixth.[3]

---

[1] Scheybl adds: Moreover this is evident by the following calculation:

$$
\begin{array}{r}
10 + 2 \\
\underline{10 - 1} \\
100 + 20 \\
\underline{- 10} \\
110 \\
\underline{- 2} \\
108
\end{array}
$$

Similarly to the example which immediately precedes this Scheybl adds: as follows:

$$
\begin{array}{r}
10 - 2 \\
\underline{10 - 1} \\
100 - 20 \\
\underline{- 10} \\
70 \\
+ 1\,[2] \\
71\,[72]
\end{array}
$$

[2] Evidently considerable interchange of text was made at this point by Scheybl. The passage inserted in the footnote to line 6 should be compared with the 28 lines of Scheybl's text on page 94, which are not found at that point in the Dresden and Vienna manuscripts.

[3] Scheybl adds: The calculation follows:

$$
\begin{array}{r}
\text{unit} - \tfrac{1}{6} \\
\underline{\text{unit} - \tfrac{1}{6}} \\
\text{unit} - \tfrac{1}{6} \\
\underline{- \tfrac{1}{6}} \\
\text{giving } \tfrac{2}{3} \\
\text{Add} \quad \tfrac{1}{6} \text{ of } \tfrac{1}{6}.
\end{array}
$$

Whence the product of this multiplication finally amounts to $\frac{2}{3}$ plus $\frac{1}{36}$.

1   Sequuntur nunc similes nodorum multiplicationes, per res seu radices et numeros expositae. Eodem modo res inter se multiplicantur, quando cum ipsis numeri vel ipsae sine numeris fuerint, vel quando sine ipsis numeri propositi fuerint. Dicendo multiplicetur 1 res et 10 cum 1 re et 10, dicas igitur res cum re substan-
5 tiam, et res cum 10 multiplicata 10 res generat. Item 10 cum re 10 res, et 10 cum 10 multiplicata 100 generant. Erit ergo totum 1 substantia 20 res et 100 ex numeris.

Et si dicas, multiplicetur 1 res cum 1 re: dicas, res cum re multiplicata producit substantiam. Atque tantum quidem est multiplicationis productum.

10   Similiter res sine 10 cum re sine 10: dic, res cum re substantiam producit; sine 10 vero cum re 10 res producit diminuendas. Item res cum re 10 res diminuendas; sine 10 verò cum sine 10 100 ex numeris addendas procreant. Vnde totum multiplicationis productum ad 1 substantiam sine 20 radicibus, additis vero ex numeris 100, sese extendit.

15   Vel etiam si dicas, 10 cum 10; item 10 sine re cum 10 sine re: dic 10 cum 10 multiplicata procreant 100. Atque tantum est productum multiplicationis prioris. Dic deinde 10 cum 10 centum addenda; sine re verò cum 10, 10 res generat diminuendas. Item dic 10 cum sine re 10 res diminuendas; sine re verò cum sine re, substantiam procreat addendam. Vnde totum multiplicationis
20 posterioris productum ad 100 absque 20 rebus, vna substantia verò adiecta, sese extendit.

Si autem quaesieris 10 et res cum suo aequali multiplicata, quantum producunt? Dic 10 cum 10, 100; et res cum 10, 10 res procreat. Item 10 cum re, 10 res; et res cum re, substantiam generat. Tota autem haec multiplicatio ad 100 ex
25 numero, 20 res et vnam substantiam sese extendet.

Quòd si sic quaesieris decem sine re cum 10, vel decem et res cum 10: producet multiplicatio prior 100 ex numeris absque 10 rebus, posterior verò 100 ex numeris et 10 res.

Si autem dixerit aliquis, decem sine re cum 10 et re multiplicata quantum
30 faciunt? Dicas 10 cum 10, 100 drachmas; et sine re cum 10, 10 res procreat diminuendas. Item 10 cum re, res 10 generant addendas; et sine re cum re, substantiam procreat diminuendam. Hoc ergo totum ad 100 drachmas proueniet, vna substantia abiecta.

Et si dixerit, decem sine re cum re: dic 10 cum re, 10 res procreant, et sine re
35 cum re, substantiam generat diminuendam. Hoc ergo ad 10 res perueniet abiecta substantia.

1–28. *om. Vide pag.* 92, *n.* 6.
7. C *add.:* vt sequitur,
     1 res   et 10
   cum 1 re.   et 10
       1 sub. et 10 res
           10 res et 100 N
Summa pro. 1 sub.   20 res 100 ex nu.
14. C *add.:* Sequitur calculus.
     1 res   sine 10
   cum 1 re.   sine 10
       1 sub. sine 10 re.
          sine 10 re. plus 100
       1 sub. sine 10(20) res plus 100

29. aliquis *om.*
30. Ducas D.   drachmas *om.*
31. et res in 10, 10 res.   generat.
32. procreant D.   Hoc *om.* D.   perueniet.
33. vna *om.* D.   C *add.:* Sequitur calculus.
       10 sine re.
   cum   10 et   re.
       100 sine 10 reb.
           10 res sine substantia
       100 sine substantia
34–36. Et . . . substantia *in marg.* D *man.* 2.
34. dicas *pro* dic *saepe.*
35. perueniat.

Similar multiplications of nodes,[1] illustrated by things or roots and numbers, follow. In the same manner the unknowns [2] are multiplied by themselves, either when numbers added to them, or when numbers are subtracted from them, or when they are to be subtracted from numbers. For example, to multiply $x + 10$ by $x + 10$, you proceed thus: x by x, x^2, and x multiplied by 10 gives $10x$; also 10 by x, $10x$, and 10 multiplied by 10 gives 100. The sum total is then x^2, $20x$, and 100.[3]

Another example: multiply x by x. You say that x by x gives x^2, and this is the product of the multiplication.

Similarly, $x - 10$ by $x - 10$: x by x gives x^2 and negative 10 by x gives negative $10x$. Also negative [4] 10 by x gives $10x$ negative, and negative 10 by negative 10 gives positive 100. Whence the total product of this multiplication amounts to x^2 less $20x$, with 100 to be added.[3]

Or also if you multiply 10 by 10, and again $10 - x$ by $10 - x$: 10 multiplied by 10 gives 100; so much is the product of the first multiplication. Then, 10 multiplied by 10 gives positive 100; negative x by 10, $10x$ negative; also 10 by negative x gives negative $10x$; negative x by negative x gives positive x^2. Whence the total product of the second multiplication extends to 100 less $20x$, with x^2 to be added.

Again, if you try to find the product of $10 + x$ multiplied by its equal you proceed thus: 10 by 10, 100; and x by 10 gives $10x$; also 10 by x, $10x$, and x by x gives x^2. The total product amounts to 100, $20x$ and x^2.

Now if you try to find the product of either $10 - x$ by 10 or $10 + x$ by 10, the first product is $100 - 10x$ and the other $100 + 10x$.

Further some one may ask, how much is the product of $10 - x$ by $10 + x$? You proceed thus: 10 by 10, 100 units, and negative x by 10 gives negative $10x$; also 10 by x gives $10x$ positive, and negative x by x gives negative x^2. This total then equals 100 units less x^2.[3]

Another problem: $10 - x$ by x. You proceed thus: 10 by x gives $10x$, and negative x by x gives x^2 to be subtracted. This then equals $10x$ less x^2.

---

[1] See p. 91, footnote 2.

[2] *Res* (literally 'thing') is used in such a technical sense that it seems better to translate by 'unknown,' as in this instance, or by x as in much of the following work.

[3] Scheybl adds to this, and to two problems below, the following calculation forms which approach the modern symbolism for the product of binomials. Scheybl prefaces with the words, 'as follows,' or 'the calculation follows':

$x + 10$	$x - 10$	$10 - x$
$x + 10$	$x - 10$	$10 + x$
$x^2 + 10x$	$x^2 - 10x$	$100 - 10x$
$10x + 100\,N$	$-10x + 100$	$+10x - x^2$
$x^2, \quad 20x, \quad 100$	$x^2 - 20x + 100$	$100 - x^2$

Scheybl uses $100\,N$ here for the number 100, following the notation employed by him in his printed works on algebra as well as in the algebra text which is found in the same manuscript with this version of Robert of Chester.

[4] In the Latin text (l. 11) Scheybl has *res cum re* instead of *sine 10 cum re*.

1   Si autem dixerit, decem et res cum re sine 10, quantum procreant? Dic 10 cum re multiplicata, 10 res generant; et res cum re, substantiam generat addendam. Item 10 cum sine 10, 100 drachmas procreant subtrahendas; et res cum sine 10 multiplicata res 10 generat diminuendas. Dicas ergo quòd haec tota 5 summa vsque ad vnam substantiam, 100 drachmis abiectis, sese extendat.

Si autem quis dixerit, decem drachmae et rei medietas cum medietate drachmae, quinque rebus abiectis, multiplicatae, quantum procreant? Dicas, decem cum medietate drachmae multiplicata, 5 drachmas procreant, et medietas rei cum medietate drachmae, quartam rei procreat addendam. Item 10 cum sine 5 rebus 10 multiplicata, 50 procreant res diminuendas. Vnde tota haec multiplicationis summa ad 5 drachmas, 49 rebus et tribus quartis vnius rei abiectis excrescet. Postea medietate rei cum sine 5 rebus multiplicata, duae substantiae et media producentur diminuendae. Tota igitur multiplicationis summa ad 5 drachmas, duabus substantiis et media nec non etiam 49 rebus et tribus rei quartis abiectis, 15 excrescet.

Et si dixerit, decem et res cum re absque 10 multiplicata, quantum faciunt? Est quasi diceres, res et 10 cum re sine 10. Vnde sic respondeas: res cum re multiplicata, substantiam generat; et 10 cum re, 10 radices generant addendas. Item res cum sine 10 multiplicata, 10 res procreat diminuendas. Vnde 10 res 20 adiectae et 10 res diminutae seu ablatae, cum prius tantum tribuat quantum posterius aufert, negligunt; et relinquitur substantia sola. Porro 10 cum sine 10, 100 drachmas generant ex omni substantia diminuendas. Tota igitur haec multiplicatio ad substantiam, 100 drachmis abiectis, extenditur.

CAPUT ADIECTIONIS ET DIMINUTIONIS

25   Sciendum est, quòd omnis radix substantiae propositae est ignota; duplicatur etiam et triplicatur et caet., atque ex ipsius duplicatione et triplicatione cum sua substantia talis nascitur numerus cuius videlicet vna radix duabus siue tribus radicibus suae substantiae aequiparatur. Quod totum euenire videtur iuxta multiplicationem numeri supra vnitatem naturaliter disposti. Nam si radices

1. Dicas res in 10.
2. generat.   res in re + ducta (deducta D).
3. et sine 10 in 10 + deducta.   diminutiua. sine (cum D) 10 in re.
4. substantia absque 100, illi (id D) cum quo opposuisti equatur (coequatur D). Quod id circo contingere videtur quia proiecisti 10 res diminutiuas cum 10 rebus (+ et D) adiectiuis. Vnde eciam substantia absque 100 dragmatibus permansit *pro* quòd haec . . . extendat.
7–10. multiplicata, quantum procreant. Dicas medietas dragmatis in 10 ducta dragmatibus, 5 dragmata progenerat (progenerant D) et medietas dragmatis in rei medietate (medietatem D) deducta, quartam rei procreat (procreant D) adiectiue, et sine 10 rebus in 10 multiplicatum dragmatibus, 50 res procreat diminutiuas. Vnde et.
11. 36 *pro* 49 *bis.*   excrescit.
12. Postea multiplica medietatem dragmatis absque 5 rebus in medietatem rei adiectiue fientque due substantie et medium diminutiue.

13. diminutiua *corr. in* diminuendae C.
17. et *pro* Est D.   diceret.   sine 10 . . . multiplicata *in marg.* D *man.* 2.
18. re + multiplicata.   10 res.   adiectiuas.
19. et sine re (x D) in re.   diminutiuas.
19–22. unde (+ re in 100 *superscr.* D *man.* 2) adiectiua cum diminutiuis adnullantur (adnullatur D) et remanet substantia, ac 10 in sine 10 100 generantur ex omni substantia diminutiua.
23. extenditur + Et quotquot (quidquid D) fuerit in multiplicatione adiectum seu diminutum semper in lance consimili apponitur id est si unum fuerit adiectiuum alterum erit diminutiuum.
24. *Titulum* Radicum algorismus V; *om.* D.
25. quam V; quando D *pro* quòd.   seu V; siue D *pro* est².   ignote.
26. aut *pro* etiam et.   ut *pro* atque.
27. substantia + multiplicatione.   nascatur.
28. seu V; siue D *pro* suae.   equiparantur D.
29. super sola unitate V; super solam unitatem D.   Nam + et.   radicem.

Yet another problem: how much is $10 + x$ by $x - 10$? 10 multiplied by x gives $10x$, and x by x gives positive x^2; also 10 by negative 10 gives 100 units negative, and x multiplied by negative 10 gives negative $10x$. You can say, therefore, that this sum total amounts to x^2 less 100 units.

If moreover some one asks what is the product of 10 units and one-half x multiplied by one-half a unit less $5x$, you proceed thus: 10 multiplied by one-half a unit gives 5 units and one-half of x by one-half a unit gives one-fourth x; also 10 by negative $5x$ gives negative $50x$. Whence the sum total of this multiplication amounts to 5 units, from which are to be subtracted[1] $49x$ and $\frac{3}{4}x$. Then $\frac{1}{2}x$ multiplied by negative $5x$ gives two and one-half x^2 negative. The sum total of the multiplication amounts to 5 units, with two and one-half x^2, and $49x$ and $\frac{3}{4}x$ to be subtracted.[2]

Another problem: how much is $10 + x$ multiplied by $x - 10$? This is the same as $x + 10$ by $x - 10$. Whence you proceed in this manner: x multiplied by x gives x^2, and 10 by x gives 10 positive roots; also x multiplied by negative 10 gives negative $10x$. Whence the $10x$ to be added (positive) and the $10x$ to be subtracted (negative), or taken away, cancel each other, since the first adds as much as the second takes away and x^2 alone remains. Then 10 by negative 10 gives 100 units to be subtracted from x^2. This total product therefore amounts to x^2 less[3] 100 units.

ON INCREASING AND DIMINISHING [4]

The fact must be recognized that every root of any given square is unknown; it is also doubled or tripled, etc. in such a way that by doubling and tripling it, by the multiplication of its square, a number is formed of which one root is equal to two or three roots of the given unknown square. All of this turns out to be like the multiplication of any number beyond unity, all in natural order. For if you wish to double the roots, you multiply

---

[1] Error made by Scheybl, who writes "added" instead of "subtracted."

[2] $\left(10 + \dfrac{x}{2}\right)\left(\frac{1}{2} - 5x\right) = 5 - 49\frac{3}{4}x - 2\frac{1}{2}x^2.$

[3] Scheybl writes *adiectis* for *abiectis*. The problem is, of course, that $x + 10$ multiplied by $x - 10$ gives $x^2 - 100$.

[4] An algebraical work with this title is supposed to have been written by Al-Khowarizmi.

The Libri and the Arabic versions follow with four problems which do not occur in Robert of Chester's translation. However, Scheybl takes up three of these problems in his additions, on pages 142–144 of this work.

These problems, following Rosen, *op. cit.*, p. 27, are as follows: "Know that the root of two hundred minus ten, added to twenty minus the root of two hundred, is just ten. The root of two hundred, minus ten, subtracted from twenty minus the root of two hundred, is thirty minus twice the root of two hundred; twice the root of two hundred is equal to the root of eight hundred. A hundred and a square minus twenty roots, added to fifty and ten roots minus two squares, is a hundred and fifty, minus a square and minus ten roots. A hundred and a square, minus twenty roots, diminished by fifty and ten roots minus two squares, is fifty dirhems and three squares minus thirty roots. I shall hereafter explain to you the reason of this by a figure, which will be annexed to this chapter."

1 duplicare volueris, binarium cum binario multiplices et quod ex multiplicatione
excreuerit, cum ipsius radicis substantia multiplices; et is excrescet numerus,
cuius vna radix duabus ipsius substantiae radicibus fiet aequalis. Quòd si radicem
triplicare volueris, ternarium cum ternario multiplices, et quod ex multiplicatione
5 excreuerit, cum ipsius radicis substantia multiplices; et is tibi nascetur numerus,
cuius vna radix tribus radicibus primae substantiae aequiparatur. Si autem
medietatem radicis habere volueris, oportet vt medietatem cum medietate, ac
cum producto postea ipsam substantiam multiplices; et erit radix quae tollitur
medietati radicis substantiae aequalis. Natura enim numeri hoc exigit, vt
10 quemadmodum in numeris integris multiplicatur, ita etiam et in numeris diminutis, hoc est in fractionibus. Eodem modo cum tertiis, cum quartis, atque omni
eo quod ipso integro minus est, agendum erit. Similitudo autem multiplicationis
huius talis est.

Similitudo multiplicationis prima. Accipiamus exempli gratia radicem numeri
15 nouenarii, ac deinde multiplicemus. Quòd si duplationem radicis numeri nouem
habere volueris, dicas, bis duo procreant 4, que cum nouenario multiplicata, ad
36 excrescet multitudo. Huius itaque multitudinis accipias radicem, id est
numerum senarium, qui duabus radicibus numeri nouenarii, hoc est numeri
ternarii duplo aequalis reperitur; idem est enim numerum senarium semel
20 accipere.

Similitudo multiplicationis secunda. Si autem radicem numeri nouem triplicare volueris, ternarium cum ternario multiplices, et fient 9; quae si cum seipsis, hoc est cum nouenario multiplicaueris, vsque ad 81 excrescet numerus, cuius
vnam radicem nouenarius complet numerus, qui tribus radicibus nouenarii, hoc
25 est numero nouem, videtur aequalis.

Similitudo multiplicationis tertia. Sed si medietatem radicis saepe dicti
numeri, habere volueris, medium cum medio multiplica, et fiet quarta, quam
si cum 9 multiplicaueris, duo et quartam vnius perficies. Horum igitur
radicem accipias, id est vnum et medium, quae medietatem radicis numeri
30 nouenarii, hoc est medietatem numeri ternarii, adimplent. Nam vnum cum sui

1. radicare *pro* duplicare D.   binario in binario D.   dicas V; ducas D *pro* multiplices. numerum ut (et D) *pro* et.

2. substantiam.   talis *pro* is.

4. ducas (+ productum in V) substantiam ut talis tibi nascatur *pro* multiplices . . . nascetur.

6. equiparantur D.

7. assumere volueris.   et *pro* ac.

8. cum producto *om.*   in *pro* ipsam.   tol-litur + hoc est elicitur C.

10. multiplico D.   diminutis + multiplicetur.

11. hoc est in fractionibus *om.*   Hoc ergo (quo- D) modo seu (siue D) in 3^{is} seu (siue D) in 4^{is} seu (siue D) in eo quod minus est.

12–14. igitur huius multiplicacione prima talis est ut *pro* autem . . . gratia.

15. Et *pro* Quòd D.   duplicationem.   nu-meri nouem habere *om.*

16. bis bini.   ducta (ducitur D) + numero.

17. multiplicatio *pro* multitudo; multitudo *corr. ex* multipli C.   summe *pro* multitudinis.

18. senarium + accipe.   unum *pro* numeri[1] D. id est numero 3^{rls}.

19. bis sumpto similis V; simul D *pro* duplo aequalis.   ternarium que est radix nouenarii, bis accipere et numerum semel accipere senarium V; senarium semel accipere quod si numerum ternarium que radix est numeri nouenarii, bis accipere D *pro* senarium semel accipere.

21. Similitudo multiplicationis secunda *om. et sic infra.*   numeri nouem *om.*

22. ducas numerum, et fiunt 9 que si (similiter D).

23. hoc est cum nouenario *om.*   cuius + vide-licet.

24. qui scilicet tribus radicibus nouenarii numeri videtur equalis (*om.* D).

26. Et *pro* Sed D.   radicis *om.* V.   saepe dicti numeri *om.*

27. multiplicare *pro* habere; multiplicare C *sed del.*   fietque.   quia *pro* quam.

29. quam *pro* quae.

30. medietatem *om.*   numeri *om.* V.

2 by 2, and the product by the unknown square of the same root. The
result will be a number of which one root will be equal to two roots of the
given unknown square.[1] And if you wish to triple the root, you multiply
3 and 3, and the product by the square of the root; so you obtain a number
of which one root is equal to three roots of the first square.[2] Moreover, if
you wish to take one-half of a root, it is necessary to multiply one-half by
one-half, and then the product by the square itself. The root which is
taken will be one-half of the root of the given square.[3] Indeed the nature
of numbers requires that just as integral numbers are multiplied so also are
lesser numbers, *i.e.* fractions. You proceed then in the same manner with
thirds, with fourths, and so with every number less than an integer; illustrations follow.[4]

First illustration[5]: Take the root of nine to be multiplied. If you wish
to double the root of nine you proceed as follows: 2 by 2 gives 4, which you
multiply by 9, giving 36. Take the root of this, *i.e.* 6, which is found to be
two roots of nine, *i.e.* the double of three. For three, the root of nine, added
to itself gives 6.[6]

Second illustration: If you wish to triple the root of 9, you multiply 3
by 3, giving 9, which multiplied by itself, *i.e.* by 9, gives 81. Of this number
9 is the root, and this is seen to be equal to 3 roots of 9, *i.e.* 3 $\times$ 3.[7]

Third illustration: If, however, you wish to take one-half of the root,
multiply one-half by one-half, giving $\frac{1}{4}$, which when multiplied by 9 will
give $2\frac{1}{4}$. Take the root of this, *i.e.* $1\frac{1}{2}$, which is one-half of the root of 9,

[1] $2\sqrt{x} = \sqrt{2^2 \cdot x}$.

[2] $3\sqrt{x} = \sqrt{3^2 \cdot x}$.

[3] $\frac{1}{2}\sqrt{x} = \sqrt{\frac{1}{2} \cdot \frac{1}{2} \cdot x} = \sqrt{\frac{x}{4}}$.

[4] This section begins in the Arabic with the four problems which we have given in footnote
4 on the preceding page of the translation. The part which corresponds to this paragraph,
following Rosen, pp. 27–28, is as follows: 'If you require to double the root of any known or
unknown square (the meaning of its duplication being that you multiply it by two), then it
will suffice to multiply two by two, and then by the square; the root of the product is equal to
twice the root of the original square.

If you require to take it thrice, you multiply three by three, and then by the square; the
root of the product is thrice the root of the original square.

Compute in this manner every multiplication of the roots whether the multiplication be
more or less than two.

If you require to find the moiety of the root of the square, you need only multiply a half by
a half, which is a quarter; and then this by the square: the root of the product will be half
the root of the first square.

Follow the same rule when you seek for a third, or a quarter of a root, or any larger or
smaller quota of it, whatever may be the denominator or the numerator. Examples of this . . .'

[5] In translating I omit some words added by Scheybl.

[6] $2\sqrt{9} = \sqrt{4 \cdot 9} = 6$. The problems may appear trivial, but the reader should note that this
is the first approach to an algebraic treatment in systematic form of surd quantities. Al-Khowarizmi proceeds admirably from known to unknown.

[7] $3\sqrt{9} = \sqrt{3 \cdot 3 \cdot 9} = 9$.

1 ipsius medio bis acceptum ternarium complet numerum. Secundum ergo hunc modum in huius modi multiplicationibus cum omnibus radicibus, quotquot integrae vel fractae fuerint, agendum erit.

Modus diuidendi

5 Si autem radicem numeri nouenarii in radicem quaternarii diuidere volueris. Diuide 9 in 4 et exeunt $2\frac{1}{4}$, quorum radix, quae in vno et dimidio terminatur, vnam complet particulam.

Si autem e contrario diuidere volueris, id est radicem numeri quaternarii in radicem numeri nouenarii, diuide 4 in 9, et exeunt quatuor nonae vnius; harum 10 radicem, id est duas tercias, accipe. Et hoc est quod vni particulae scilicet contingit.

Alius diuisionis modus

Quòd si radicem numeri nouenarii in radicem numeri quaternarii diuidere volueris, ita tamen vt substantia in substantiam non diuidatur, radicem numeri 15 nouenarii, quoties volueris, duplica vel collige, et scias cuius numeri numerus ex collectione proueniens radix habeatur. Hunc ergo modum numeri in 4, aut in alium numerum, in quèm radicem primam diuidere voluisti diuide. Nam eius radix vni eueniet. Iuxta ergo hunc modum si tres radices vel quatuor, seu pauciores, seu medietatem radicis, aut minus, aut quotquot fuerint, numeri 20 nouenarii diuidere volueris, cum omnibus iis agendum erit. Operare ergo secun · dum modum quem proposuimus, et rem ita se habere inuenies, si deus voluerit.

Sequuntur nunc similes multiplicationes, per res seu radices et numeros expositae

Sed si radicem numeri nouenarii cum radice numeri quaternarii multiplicare volueris, multiplicemus 9 cum 4 et producentur 36. Sume ergo horum radicem, 25 id est 6. Et hoc est quod producitur ex radice numeri nouenarii cum radice numeri quaternarii multiplicata. Eodem modo si multiplicare volueris radicem numeri quinarii cum radice numeri denarii. Multiplica ergo 5 cum 10, et producentur 50, quorum radix substantiam, hoc est ipsum quod voluisti, significat.

1. mediabis *pro* medio bis D.

2. modi *om.* D.   multiplicatione quotquot radices seu (*om.* D) integre seu diminute fuerint.

4. *Titulum om. et sic infra.*

5. radices *pro* radicem [1] D.   numeri *om.* V. numeri quaternarii D.   diuide per *vel* diuide super *pro* diuide in *saepe.*

6. Deinde *pro* Diuide D.   fientque duo et 4ᵃ. quarum D.   quae + scilicet radix.

8. econuerso.   super *pro* in.

9. numeri *om.* D.   fientque.   nouene D. tunc earum radicem assume id est duas unius accipe 3ᵃˢ *pro* vnius . . . accipe.

10. particulae scilicet *om.*

13. super.

14. tamen quod substantiam super.

15. quotiens.   duplica V; multiplica duplica D.   vel collige *om.*   scito.

16. ex duplicatione concretus radix habeatur. Per hunc.

17. primo V; primum D.

18. Nam + et.   ergo *om.* D.   aut plures *pro* vel quatuor.

19. aut quotquot fuerint, numeri nouenarii *om.* V; aut quod volueris, numeri nouenarii D; quotquot + volueris C *sed del.*

20. cum omnibus iis *om.*   est.   Operacio D. secundum quod diximus et inuenies si deus voluerit.

23. Sed + et.   radicem *om.* D.   numeri [2] *om.*

24. Multiplica.   fientque 36.

25. quod excreuit.

26. multiplica D.   numeri *om.* D.

27. numeri *om.*   senarii D.   ergo *om.*

28. quorum videlicet radix substantiam quam voluisti signat.

i.e. one-half of 3. For $1\frac{1}{2}$ taken twice gives 3.[1] You proceed then in the same manner with such multiplications with all roots, whether they are integral or fractional.

Method of dividing

If, moreover, you wish to divide the root of 9 by the root of 4, you divide 9 by 4, giving $2\frac{1}{4}$, and the root of this, which is finally $1\frac{1}{2}$, completes the division.[2]

If you wish to perform the reciprocal division, *i.e.* divide the root of 4 by the root of 9, divide 4 by 9, giving $\frac{4}{9}$ of a unit, and take the root of this, *i.e.* $\frac{2}{3}$. This is, of course, the result of the division.[3]

Another method of division

You may desire to divide the root of 9 by the root of 4, without dividing the square by the square.[4] Double or gather up the root of 9 as many times as desired and of the resulting number you find the root. This number divide by 4 or by any other number by which you wished to divide the first root; for in this way the root of it will be found. You proceed then in like manner if you wish to divide the root of 9 by 3 or 4 or less, or by $\frac{1}{2}$ or less, or by anything else; with all of these the rule is the same. Follow then the rule which we have explained and so you will find the result, if God will.

Similar multiplications explained by things, or roots, and numbers

Now if you wish to multiply the root of 9 by the root of 4, multiply 9 by 4, giving 36. Take the root of this, *i.e.* 6. This is the product of the root of 9 multiplied by the root of 4. Likewise if you wish to multiply the root of 5 by the root of 10, you multiply 5 by 10 giving 50.[5] The root of this is the desired product.

[1] $\frac{1}{2}\sqrt{9} = \sqrt{\frac{9}{4}} = \frac{3}{2}$.

[2] $\dfrac{\sqrt{9}}{\sqrt{4}} = \sqrt{\frac{9}{4}}$.

[3] $\dfrac{\sqrt{4}}{\sqrt{9}} = \sqrt{\frac{4}{9}} = \frac{2}{3}$.

[4] There seems to be something incorrect about Robert's translation. Probably this should be as in the Arabic, according to Rosen, *op. cit.*, p. 30: "If you wish to divide twice the root of 9 by the root of 4, or of any other square, you double the root of nine in the manner above shown to you in the chapter on Multiplication, and you divide the product by four, or by any number whatever. You perform this in the way pointed out.

In like manner, if you wish to divide three roots of nine, or more, or one-half or any multiple or sub-multiple of the root of nine, the rule is always the same: follow it, the result will be right."

Rosen here follows the custom of modern translators of Arabic in leaving out the reference to the Deity which is actually given in the Arabic text.

[5] $\sqrt{5} \cdot \sqrt{10} = \sqrt{50}$, both quantities being surds.

1 Quòd si radicem tertiae cum radice medietatis multiplicare volueris, tertiam cum medietate multiplica, et producitur vna sexta. Radix igitur sextae ipsum est quod ex radice tertiae cum radice medietatis multiplicata, nobis excreuit.

Si autem duas radices numeri nouenarii, cum tribus radicibus numeri quater-
5 narii multiplicare volueris, accipe duas radices numeri nouenarii secundum quod iam diximus, vt scias cuius substantiae radicem compleant. Similiter de tribus radicibus numeri quaternarii facias, quatenus cuius substantiae sint radix reperias. Has igitur substantias inter se multiplica, vnam videlicet earum cum altera, atque huius producti radicem accipe, quoniam hoc est quod ex duabus
10 radicibus numeri nouenarii cum tribus radicibus quaternarii multiplicatis excreuit.

Quotcunque igitur radices simul colligere, vel quas a quibusdam minuere volueris, cum errore abiecto iuxta hoc exemplar multiplicare poteris.

INCIPIUNT CAPITUM QUAESTIONES

Dixit Mahomet Algoarizim, hactenus praemisimus numerorum capita, quae sub
15 sex quaestionibus pro numero capitum in libri principio a nobis proposita sunt atque ibi etiam diximus, numerum restaurationis et oppositionis in his sex capitibus omnino versari. Sed quoniam ea quasi sub inuolucris te edicta sunt, igitur haec ipsa, quò omnium studium exerceatur, et scientia facilius elucescat, adducemus ac fusius explicamus.

20   *Caput primum, quaestio prima*

Modus huius quaestionis est, vt dicas: denarium numerum in duo diuide vt eius vna pars cum altera multiplicata, numerum ex multiplicatione concreet seu producat, qui quater acceptus aequalis fuit numero, ex multiplicatione vnius partis semel cum se ipsa generato.

25 Similitudo talis est, vt vnam partem numeri denarii rem constituas, et alteram 10 sine re. Multiplica igitur rem cum 10 sine re, fient 10 res absque substantia. Item multiplica 10 res absque substantia cum quatuor, quoniam quater dixisti,

2. fietque $\frac{1}{8}$ (6ᵃ D).   igitur *om.* D.   ipsa.

4. cum tribus radicibus numeri quaternarii *om.* D.

5. extrahe.

7. fuit D.

8. inuicem *pro* inter se.   et unam (id est una D) earum in alteram ducas, et huius summe radicem accipias.

9. erit.

10. radicibus *om.*   numeri 4ʳⁱⁱ.

11. Quotquot.   in simul colligere volueris. inuicem *sed del. et* minuere *superscr.* D *man.* 2.

12. omni *pro* cum.   poteris + laus deo etc. V; poteris *om.* D.

13. *Titulum om.*

14–16. Mahumed Algoarizim, primum quidem capitula numeri proposuimus (premisimus D) que sub sex questionibus ad similitudinem 6 capitulorum in libri principio propositorum constituimus. Vbi eciam diximus quam numerus.

17. procul dubio versatur.

17–19. ea que in questionibus quasi sub quodam

tegumento sunt dicta, hic illud in quo sciencia et animi studium faciliori exerceantur aditu elucescet introducimus.   C *add.:* Quid si quaestio capitis primi?   te *pro* tibi C.

20. *Titulum om. et sic ubique.*

21. *In marg.* C Textus, *et sic ubique in has quaest.* Modus huius capituli.   decenum *pro* denarium *fere ubique.*   ita in duo *et sic saepe.* diuido.

22. diuisio *pro* pars *vel* pars diuisionis *saepius.*

22–24. deducta, numerus ex multiplicatione increet et quater acceptus similis sit numero (non D) ex multiplicatione unius diuisionis semel in seipsam deducte generato.

25. *In marg.* C Minor, *et sic ubique.*   decem *pro* denarii.   alteram + eiusdem (eius D) diuisionem.

26. re¹ + proponas V; + propones D.   re² *om.* V.   fientque 10 radices.

27. 10 sine re V; rem D *pro* 10 res absque substantia.

If you wish to multiply the root of $\frac{1}{3}$ by the root of $\frac{1}{2}$, you multiply $\frac{1}{3}$ by $\frac{1}{2}$, giving $\frac{1}{6}$. The root of this one-sixth is that which we obtain by the multiplication of the root of $\frac{1}{3}$ by the root of $\frac{1}{2}$.

Again if you wish to multiply two roots of 9 by three roots of 4,[1] take two roots of 9 according to the method which we have explained so that you may know the square of which this is the root. Treat similarly the three roots of 4 in order to find the square of which this is the root. Then multiply these squares by each other, *i.e.* one of them by the other, and take the root of this product since this is the result of the multiplication of two roots of 9 by three roots of 4.

Therefore using this process you are able, casting aside error, to multiply as many roots as you wish to join together or as many as you wish to subtract from other quantities.[2]

PROBLEMS ILLUSTRATING THE CHAPTERS [3]

Says Mohammed Al-Khowarizmi: Up to this point we have set forth the chapters on numbers which we proposed in the beginning of this work, under six problems, one for each chapter, and we also mentioned in that place that every problem of restoration and opposition necessarily falls within the scope of one of these six chapters. But since the explanations were somewhat involved we present and more fully explain these further problems, by which each type is illustrated and the science is more easily elucidated.[4]

First problem, illustrating the first chapter

Divide ten into two parts in such a way that one part multiplied by the other and the product, or result, taken four times, will be equal to the product of one part by itself.[5]

The method is to let x represent one part of ten, and the other $10 - x$. Therefore multiply x by $10 - x$, giving $10x - x^2$. Also multiply $10x - x^2$ by 4, as it was to be taken four times, giving $40x - 4x^2$ as four times the

---

[1] $2\sqrt{9}$ by $3\sqrt{4} = \sqrt{36}$ by $\sqrt{36} = \sqrt{1296} = 36$.

[2] Rosen, *op. cit.*, p. 31, translates this paragraph: "You proceed in this manner with all positive or negative roots"; he follows with the geometrical explanation of the two problems given in footnote 4, p. 33, and a further elucidation of the third problem of that set.

[3] Scheybl adds to the statement of each problem the marginal word, *Textus*, and to the explanation, *Minor;* also the numbering of the problems appears to be his addition.

[4] Rosen, *op. cit.*, p. 35: "Of the six problems.

Before the chapters on computation and the several species thereof, I shall now introduce six problems, as instances of the six cases treated of in the beginning of this work. I have shown that three among these cases, in order to be solved, do not require that the roots be halved, and I have also mentioned that the calculating by completion and reduction must always necessarily lead you to one of these cases. I now subjoin these problems, which will serve to bring the subject nearer to the understanding, to render its comprehension easier, and to make the arguments more perspicuous."

[5] $4x(10 - x) = x^2$; $5x^2 = 40x$; $x = 8$.

1 fientque quatuor aequales multiplicationes partis vnius cum altera, 40 res absque
4 substantiis. Postea rem cum re, id est alteram partem cum seipsa, multiplica,
et producetur substantia, rebus 40 absque 4 substantiis aequalis. Restaura
igitur numerum, hoc est substantiae substantias quatuor adiicias, venientque
5 quinque substantiae 40 res coaequantes. Vnius ergo substantiae radicem octonarius numerus assignat. Ipsa deinde substantia in 64 terminatur, cuius scilicet
radix vnam partem numeri denarii cum seipsa multiplicatam, demonstrat, et
residuum numeri 10 in binario terminatur numero. Duo itaque alteram partem
numeri 10 obtinent. Iam igitur haec quaestio ad vnum sex illorum capitum,
10 illud nimirum in quo diximus, substantiae radices coaequant, te perduxit.

Caput secundum, quaestio secunda

Denarium numerum sic in duo diuido, vt si denarius numerus semel cum seipso
multiplicetur, numerus qui ex multiplicatione producetur aequalis sit duplo
numeri eius, qui ex multiplicatione vnius partis cum seipsa producitur, septem
15 nouenis partibus eiusdem producti numeri superadditis.

Huius rei expositio est, vt numerum 10 cum seipso multiplices, fientque 100,
duas substantias et septem vnius substantiae nonas coaequantia. Haec ad vnam
conuertas substantiam, id est ad nouem vigesimas quintas partes quae quintam
et $\frac{4}{5}$ quintae continent, atque centenum numerum ad eius quintam et quatuor
20 quintae quintas, id est ad 36, vnam substantiam coaequantes conuerte; et erit
radix substantiae 6, vnam diuisionis partem numeri decem exprimens, vnde
altera deinde in quaternario numero procul dubio terminatur. Igitur haec
quaestio ad secundum sex capitum te perduxit, in quo diximus, substantiae
numeros coaequant.

25 Caput tertium, quaestio tertia

Denarium numerum ita in duo diuido, vt, vna eius parte in alteram diuisa, inde
exiens particula in quarternario numero finietur.

Expositio talis est, vt vnam partem, rem constituas; atque alteram, 10 sine re
proponas, deinde 10 sine re in rem diuidas, et exeunt 4. Iam autem manifestum
30 est, si id quod ex aliquo diuiso exierit, cum eo in quod ipsum diuiditur, multipli-

---

1. similitudines multiplicationis.   in alteram + et erunt.

2. re + multiplica.   deducas.

3. fiet.   4 + 4^{or} D.   coequales V.

4. igitur + vel comple C.   et super sub-stantiam, substantias (res D) adicias, fientque 40 res 5 substantias.

6. et ipsa.   deinde om.   8 pro scilicet V.

8. ternario D.   Nam et (om. D) duo.

9. obtinebit C.   ad unum capitulorum te perduxit in quo diximus, Substantia radices coequat.

13-15. ducatur, numerus qui ex multiplicatione excreuerit (decreuerit D) similis sit numero qui ex multiplicatione unius diuisionis in semetipsam bis deducte (+ tollitur D) vii nouenis superadditis.

16. semetipso V.   -que om.

17. nouenas.   coequantur D.   Haec + igitur.

18. 9 partes ex (et D) 25 que (+ numerum D).

19. 4^{or} 5te quintas.   continent, et centenum numerum ad eius quintam V; om. D.

20-22. Radix igitur huius (om. D) substantie id est 6 unum numeri 10 diuisionem ostendunt. Altera uero eius diuisio in 4^{rlo}.

22-24. Igitur hec questio quo ad unum vi capitulorum usque perduxit in quo diximus, substantia numeri coequat D; om. V.

26. et eiuus (?) una D.   4 eueniant V; Vna-quoque particula in ternario numero finiatur D pro inde . . . finietur.

29. ergo pro deinde.   super rem diuide ut fiant 4. Et iam.   autem om.

30. quam pro si.   exierit + si V.   multipli-catur V.

product of one part by the other. Then multiply x by x, $i.e.$ one part by itself, giving x^2, which equals $40x - 4x^2$. Therefore restore [1] or complete the number, $i.e.$ add four squares to one square, and you obtain five squares equal to $40x$. Hence 8 is the root of the square which itself is 64. The root of this is that part of 10 which is to be multiplied by itself, and the difference between this number and ten is 2. So that 2 is the other part of 10. Now this problem has led you to one of the six chapters [2] and, indeed, to that one in which we treat the type, squares equal to roots.

Second problem, illustrating the second chapter

I divide 10 into two parts in such a way that the product of 10 by itself is equal to twice the product of one part by itself, adding seven-ninths of the same product.[3]

Explanation. You multiply 10 by itself, giving 100, which is equal to $2x^2$ and $\frac{7}{9}x^2$. You reduce this to one square, $i.e.$ to $\frac{9}{25}$, equal to $\frac{1}{5}$ and $\frac{4}{5}$ of $\frac{1}{5}$ of itself. And so reduce $\frac{1}{5}$ and $\frac{4}{5}$ of $\frac{1}{5}$ of 100, $i.e.$ 36, to one square. The root of the square is 6, representing one part of the division of 10, whence then the other part is necessarily 4. Therefore this problem has led you to the second of the six chapters in which we treated the type, squares equal to numbers.

Third problem, illustrating the third chapter

I divide 10 into two parts in such a way that when one part is divided by the other, the resulting fraction equals 4.[4]

Explanation. You let x represent one part, and consequently the other you propose as $10 - x$. Then you divide $10 - x$ by x, giving 4. Now it is evident that if you multiply the quotient by the divisor you

---

[1] This is strictly the operation corresponding to the term $algebra$, and the verb used in the Arabic text is in fact from the same stem jbr as the word 'algebra'; the quantity $40x$ above is regarded as incomplete by the amount $4x^2$.

[2] The reference is to the six types of quadratic equations which are discussed extensively in the first part of the work, namely,

$$ax^2 = bx, \qquad ax^2 + bx = n,$$
$$ax^2 = n, \qquad ax^2 + n = bx,$$
$$ax = n, \qquad ax^2 = bx + n.$$

The first six of the problems in this set are chosen to illustrate each of these six types, in order.

[3] $2x^2 + \frac{7}{9}x^2 = 100$; $\frac{25}{9}x^2 = 100$; $x^2 = 36$ and $x = 6$.

The Arabic text of this problem is somewhat lengthier, and includes the statement of a second problem which does not appear either in our text or in the Libri version. Following Rosen's translation, $op.$ $cit.$, p. 36: "I have divided ten into two portions: I have multiplied each of the parts by itself, and afterwards ten by itself: the product of ten by itself is equal to one of the two parts multiplied by itself, and afterwards by two and seven-ninths; or equal to the other multiplied by itself and afterwards by six and one-fourth." But a solution is given only for the first part of the problem.

[4] $\frac{10 - x}{x} = 4$; $10 - x = 4x$; $10 = 5x$, and $x = 2$.

ı caris, quòd tum substantia quae diuisa est, adimpleatur. Et in hac quaestione
4 diuisum obtinet et id per quod diuisum diuiditur, rem proposuimus; numerus
igitur 4 cum re multiplicandus erit, et producentur 4 res, substantiam quam
diuisimus, hoc est 10 sine re exequantes. Restaura igitur 10 sine re, et ipsam rem
5 rebus 4 adde, et venient 10 quinque res coaequantia. Haec autem res in binario
finitur numero. Vnde haec quaestio ad tertium sex capitum te perduxit, vbi etiam
diximus, radices numeros coaequant.

Caput quartum, quaestio quarta

Tertiam rei et vnam drachmam cum quarta rei et vna drachma sic multiplico,
10 vt huius multiplicationis productum in 20 terminetur.

Expositio talis est, vt tertiam rei cum quarta rei multiplices, et producetur
medietas sextae vnius substantiae, et drachma cum quarta rei multiplicata,
quartam rei generat addendam. Similiter tertia rei cum drachma, tertiam
rei procreat atque tandem drachma cum drachma, drachmam producit. Haec
15 porro multiplicatio ad medietatem sextae vnius substantiae, et ad tertiam ac
quartam rei atque ad vnam drachmam vinginti drachmas coaequantia extenditur.
Vnam igitur drachmam ex 20 subtrahas, et manebunt 19 drachmae medietatem
sextae vnius substantiae simul tertiam ac quartam rei coaequantes. Iam ergo
substantiam compleas, hoc est, quicquid habueris cum duodecim multiplices, et
20 producetur substantia et 7 res, 228 ex numeris coaequantes. Radices igitur
media, id est per medium diuide, et vnam medietatem cum seipsa multiplica, et
producentur 12 et $\frac{1}{4}$. Haec 228 ex numeris adiicias, et veniunt 240 et $\frac{1}{4}$; hinc
radicem accipe, 15 et $\frac{1}{2}$; atque ex ea 3 et dimidium subtrahe, et manebunt 12,
radicem substantiae adimplentia. Iam igitur haec quaestio ad quartum sex
25 capitum te perduxit, in quo diximus, substantia et radices numeros coaequant.

Caput quintum, quaestio quinta

Denarium numerum ita in duo diuido, vt vnaquaque diuisionis parte cum
seipsa multiplicata, duorum productorum summa ad 58 perueniat.

1. substantiam cuius diuisiones (diuisionis D) fuerint adimplebit *pro* quod . . . adimpleatur. hac diuisione D.

2. cum quatuor, ipsum exeuntem, id deinde in quod diuiditur, rem proponamus C. Multiplica ergo rem per 4 et fient 4 res V; Multiplicatam ergo iiiior et fient xi res D.

4. coequantes.   cum re D *pro* sine re.   et super 4 res ipsam adde, fientque.

5. ergo.

6. Ergo *pro* Vnde.   usque ad unum sex. ut et V; nos (*sed del.*) ubi et D *pro* vbi etiam.

9. Terciam substantie et unum dragma in 4am substantie et uno dragmate.

10. summa *pro* productum *ubique.*

11. fietque.

12–15. in dragmate ductum, dragma generat (generatur D) adiectiuum et 3a rei in dragmate ducta 3am rei procreat. Sed et 4a rei in dragmate

ducta, 4am (4te D) rei progenerat. Hec igitur summa.

15. et rei 4am et etiam.

16. drachmas *om.*

18. sextam D.   et 3am et.   coequantia V; coe + adequantia D.   Sic igitur substantiam compleas et quiddid habueris.

20. 7 radices.   numero V; *om.* D *sed spat. relict.*

21. in ipsam.   fientque.

22. Hec igitur super 228 adicias et fient.   Huius (Hanc D) ergo.

23. et ex eis tria et medium diminuas.

24. Hec ergo V; Iam in hoc D.   ad unum 6.

25. substantie.   numerum.

27. sic *pro* ita.   *In marg.* |$\phi + 3\,\mathfrak{R}$| D.

unaquaque diuisione in semetipsam deducta, tota multiplicationis summa ad 58 tendatur.

obtain thus the quantity[1] which was divided. And as in this problem the quotient is 4 and the divisor is given as x, you multiply 4 by x, giving $4x$ for the quantity which was divided, *i.e.* equal to $10 - x$. Therefore complete $10 - x$ by adding x to $4x$, giving 10 equal to $5x$. Whence it follows that x is 2. Thus this problem has led you to the third of the six chapters in which we treated the type, roots equal to numbers.

Fourth problem, illustrating the fourth chapter

Multiply $\frac{1}{3}x$ and one unit by $\frac{1}{4}x$ and one unit so as to give as the product 20.[2]

Explanation. You multiply $\frac{1}{3}x$ by $\frac{1}{4}x$, giving $\frac{1}{2}$ of $\frac{1}{6}x^2$, and a unit multiplied by $\frac{1}{4}x$ gives $\frac{1}{4}x$ to be added. Similarly $\frac{1}{3}x$ multiplied by a unit gives $\frac{1}{3}x$ and then a unit by a unit gives a unit. Then this multiplication amounts to $\frac{1}{2}$ of $\frac{1}{6}x^2$, and $\frac{1}{3}x$ and $\frac{1}{4}x$ and one a unit, equal to 20 units. You subtract one unit from 20 units, giving 19 units equal to $\frac{1}{2}$ of $\frac{1}{6}x^2$ together with $\frac{1}{3}x$ and $\frac{1}{4}x$. Now then you complete the square,[3] *i.e.* you multiply throughout by 12. This gives x^2 and $7x$ equal to 228. Then halve the roots, *i.e.* divide them equally, and multiply one-half by itself, giving $12\frac{1}{4}$. You add this to 228, and you will have $240\frac{1}{4}$. From the root of this, $15\frac{1}{2}$, subtract $3\frac{1}{2}$, leaving 12 as the root of the square. Now then this problem has led you to the fourth of the six chapters in which we treated the type, a square and roots equal to numbers.

Fifth problem, illustrating the fifth chapter

I divide ten into two parts in such a way that the sum of the products obtained by multiplying each part by itself is equal to 58.[4]

[1] Robert of Chester employs *substantia* here in the non-technical sense of "substance" or "quantity." So also *census* in the Libri version and *mal* in the Arabic version are used with the same significance. This was a common usage of Arabic writers. Abu Kamil followed this practice, on occasion, and Leonard of Pisa, who drew extensively from Abu Kamil, copied this peculiarity from the Arabs: see *Scritti di Leonardo Pisano*, Vol. I, p. 422, and my article, *The Algebra of Abu Kamil Shoja' ben Aslam, Bibliotheca Mathematica*, third series, Vol. XII (1912–1913), p. 53. In particular, also, *census* appears in this sense in the *Liber augmenti et diminutionis vocatus numeratio divinationis, ex eo quod sapientes Indi posuerunt, quem Abraham compilavit et secundum librum qui Indorum dictus est composuit*, published by Libri, *Histoire des sciences mathématiques en Italie*, Vol. I, Paris, 1838, pp. 304–371. This work is probably by Abraham ibn Esra.

[2] $(\frac{1}{3}x + 1)(\frac{1}{4}x + 1) = 20$; $\frac{1}{12}x^2 + \frac{1}{3}x + \frac{1}{4}x + 1 = 20$; $x^2 + 7x + 12 = 240$; $x^2 + 7x = 228$; $\frac{1}{2}$ of 7 is $3\frac{1}{2}$, $(3\frac{1}{2})^2 = 12\frac{1}{4}$, $228 + 12\frac{1}{4} = 240\frac{1}{4}$, $\sqrt{240\frac{1}{4}} = 15\frac{1}{2}$, $15\frac{1}{2} - 3\frac{1}{2} = 12$, which is the value of x.

[3] The present usage of the expression "to complete the square" is quite different from that of our text. Here it means, of course, to make the coefficient of x^2 unity and this also corresponds to the operation termed by the Arabs, *algebra*, as opposed to the operation of *almuqabala;* see the article *al-Djabr wā-'l-Mukābala* by Professor H. Suter in *The Encyclopedia of Islam*, Vol. I (Leyden, 1913), pp. 989–990.

[4] $x^2 + (10 - x)^2 = 58$; $2x^2 - 20x + 100 = 58$; $2x^2 + 42 = 20x$; $x^2 + 21 = 10x$, which is a problem that appears earlier in the text.

1   Expositio talis est, vt 10 sine re cum seipso multiplices, fientque 100 et substantia
absque 20 rebus. Postea multiplica rem cum suo aequali et producetur substantia.
Deinde haec duo multiplicationis producta in vnum collige, et veniunt 100 et 2
substantiae absque 20 rebus, 58 coaequantes. Comple igitur 100 et 2 substantias
5 absque 20 rebus, et ipsas 20 res 58 ex numeris adiicias, et venient 100 drachmae et
2 substantiae, 58 drachmas et 20 res coaequantia. Hoc iam ad vnam conuertas
substantiam, atque oppositione deinde 29 ex 50 proiicias, et manebunt 21 et substantia, 10 res coaequantia. Res igitur mediabis, et veniunt 5 ; haec sum seipsis
multiplicata, producunt 25. Ex his 21 abiicias, et relinquentur 4. Accipe horum
10 radicem 2, atque hanc a 5, id est à medietate radicis, subtrahe, et manebunt 3
quae videlicet vnam partem numeri decem adimplent. Igitur haec quaestio ad
quintum sex capitum te perduxit, in quo diximus, substantia et numeri radices
coaequant.

Caput sextum, quaestio sexta

15   Rei tertiam et eius quartam sic multiplico, vt multiplicationis productum ipsam
rem, viginti quatuor drachmis superadditis, coaequent.

Expositio est, vt primum scias, quòd quando tertiam rei cum quarta rei multiplicaueris, medietas sextae vnius substantiae, rem et 24 drachmas coaequans,
oriatur. Multiplicatio igitur medietatis sextae substantiae cum duodecim, sub-
20 stantiam reddet completam. Similiter multiplicatio rei et 24 drachmarum cum
12, radices 12 ducenta et 88, substantiam coaequantes, adimplebit. Diuide
igitur radices per medium, et mediam partem cum seipsa multiplica, atque multiplicationis productum numero 288 iunge, et venient 324. Horum nunc radicem
accipe, id est 18, quibus medietatem radicum etiam adde, et radix in 24 finietur.
25 Haec igitur quaestio ad sextum sex capitum iam te perduxit, in quo diximus,
radices et numeri substantiam coaequant.

Ad huc restat, vt de sedecim aliis tractemus quaestionibus, quae ex sex praemissis oriri videntur, vt quicquid ex numero huic arti addicto opifici propositum
fuerit, omni errore abiecto facilius elici queat.

1. in semetipsis.

2. radicibus.

3-10. has multiplicationis summas in unum collige et habebis 100 et duas substantias absque 20 radicibus 58 coequantes. Comple igitur 100 et duas substantias absque 20 radicibus cum re quam diximus, et adde eam super 58, et fient 100 et due substantie, 58 et 20 (10 D) res coequancia. Res ergo mediabis et erunt 5. Hec igitur in seipsis (semetipsis D) multiplica et erunt 25. Ex his ergo 21 (*om.* D) abicias et remanebunt 4. Sume ergo horum radicem (harum radices D) id (*om.* D) est (*om.* D) duo qui ab 5 prius positis, id est a medietate radicum, subtrahas.

11. videlicet *om.*

12. unum 6.   te + iam.   numeri *om.* D. R¹ *pro* radices.

15. Substantie *pro* Rei.   in *pro* et.   et *pro* vt *et sic infra* (3).   *In marg.* | φ + ℞ 3 D.

16. substantiam.   coequant.

18. et *om.* D.

19. orietur.

20. Similiter + et.   34 V.   in (+ in D) 12.

21. ducentos et 88 et radices xii D.   coequans V ; *om.* D.

22. ac *pro* atque.

23. summam super 288 adicias (addicias D *et sic saepius*) et erit hoc totum 324 ; erit hoc totum C *sed del.*   igitur *pro* nunc.

24. accipias.   quibus videlicet medietatem radicum adicias, id est 6, et sic substantia (substantiam D) in 24 (cxiiii D) terminetur.

25. unum sex.

27. Hee ergo sunt (+ vi D) questiones de quibus superius me tractaturum promisi. Sed adhuc *pro* Ad huc.   primis *pro* praemissis.

28. quidquid V ; quid D.   huius artis intento.

29. abiecte D.   eliciatur *pro* elici queat.

Explanation. You multiply $10 - x$ by itself, giving 100 and x^2 less $20x$. Then multiply x by itself, giving x^2. Collecting the products of these two multiplications you obtain 100 and $2x^2$ less $20x$ equal to 58. Complete the 100 plus $2x^2$ less $20x$ by adding the $20x$ to 58. This gives 100 units $+ 2x^2$ equal to 58 units and $20x$. Now you reduce this to one square, and then by opposition[1] you take 29 from 50, leaving $21 + x^2$ equal to $10x$. Therefore you halve the roots, giving 5. Multiply this by itself, giving 25. From this you subtract 21 and 4 is left. Take the root of this, 2, and subtract it from 5, *i.e.* from the half of the roots. This gives 3 which represents, of course, one part of ten. So this problem has led you to the fifth of the six chapters in which we treated the type, a square and numbers equal to roots.

Sixth problem, illustrating the sixth chapter

I multiply $\frac{1}{3}x$ and $\frac{1}{4}x$ in such a way as to give x itself plus 24 units.[2]

Explanation: first you observe that when you multiply $\frac{1}{3}x$ by $\frac{1}{4}x$ you obtain $\frac{1}{2}$ of $\frac{1}{6}x^2$ equal to $x + 24$ units. The multiplication of $\frac{1}{2}$ of $\frac{1}{6}x^2$ by 12 gives the complete square. Similarly, the multiplication of $x + 24$ units gives $12x + 288$[3] which equal x^2. Therefore take one-half of the roots and multiply the half by itself. Add the product of this multiplication to 288, giving 324. Take now the root of this, *i.e.* 18, and add it to the half of the roots. The root finally is 24. So this problem has led you to the sixth of the six chapters in which we treated the type, roots and numbers equal to a square.[4]

SIXTEEN ADDITIONAL PROBLEMS [5]

It now remains for us to treat sixteen other problems which seem to arise out of the six which we have set forth, in order that the craftsman versed in this art may more easily and without any error solve any problem proposed.[6]

---

[1] This operation corresponds to the Arabic term *almuqabala*. In this instance the 29 on the right balances or cancels an equal amount of the 50 on the left.

[2] $\frac{1}{3}x \cdot \frac{1}{4}x = x + 24$; $x^2 = 12x + 288$; $\frac{1}{2}$ of 12 is 6, $6^2 = 36$, $288 + 36 = 324$, $\sqrt{324} = 18$, $18 + 6 = 24$, which is the value of the unknown.

[3] Evidently Robert of Chester wrote the word for two-hundred out in full, or in Roman numerals, and the 88 in Hindu-Arabic numerals, for both the Dresden and Scheybl versions use this form, separating the two-hundred from the 88.

[4] The above six problems appear in this order in all of the versions.

The terminology of the fourth and sixth problems is the same in the Arabic text, although Rosen, *op. cit.*, pp. 38, 40 translates differently the same expression: "I have multiplied one-third of thing and one dirhem by one-fourth of thing . . ."; "I have multiplied one-third of a root by one-fourth of a root . . ." The Libri text has *census* in both problems, and the Boncompagni text has *multitudo;* the Arabic word is *mal* with the meaning (footnote 1, p. 107) "unknown quantity."

[5] I have added this title to correspond to the Arabic (Rosen, *op. cit.*, p. 41).

[6] This paragraph is an addition by Robert of Chester.

Quaestio prima

Denarium numerum sic in duo diuido, vt vna parte cum altera multiplicata,
productum multiplicationis in 21 terminetur.

Iam ergo vnam partem, rem proponimus quam cum 10 sine re, quae alteram
5 partem habent, multiplicamus, et producuntur 10 res absque substantia, drachmas
21 coaequantes. Comple igitur 10 res cum substantia, et substantiam numero
21 adde, et venient 10 res, substantiam et 21 drachmas coaequantes. Accipe
medietatem rerum, hoc est 5, et eam cum seipsa multiplica, et producentur 25.
Ex his 21 subtrahe, et manebunt 4. Horum radicem, 2 scilicet, accipe, atque
10 eam tandem ex medietate radicum subtrahe, et manebunt 3, quae vnam partem
diuisionis demonstrant.

Quaestio secunda

Numerum denarium sic in duo diuido, vt vtraque parte cum seipsa multiplicata,
si productum partis minoris ex producto partis maioris auferatur, quadraginta
15 maneant.

Exempli expositio talis est, vt 10 sine re cum suo aequali multiplices, et producentur 100 ex numero, vna substantia absque 20 rebus. Multiplices etiam
rem cum se, et producetur substantia, quam ex 100 et substantia absque rebus
diminuas, et manebunt 100 absque 20 rebus, quadraginta drachmas coaequantia.
20 Comple igitur 100 drachmas cum 20 rebus, et eas drachmis 40 adiicias et habebis
100, quadraginta drachmas et 20 res coaequantes. Igitur 40 ex 100 auferas, et
manebunt 60 drachmae 20 res aequantes. Res igitur ternario aequantur numero,
qui vnam partem diuisionis demonstrat.

Quaestio tertia

25 Denarium numerum sic in duo diuido, vt vtraque parte cum seipsa multiplicata,
et multiplicationum productis simul collectis, ac quantitate deinde, quae est inter
duas partes, illis addita, tota summa ad 54 drachmas excrescat.

Huius exempli expositio talis est, vt 10 sine re cum suo aequali multiplices, et

---

2. vt *om.* D.   deducta, summa.   *In marg.*

$\lfloor 3 + \phi$ D.

5. fiuntque.   drachmas *om.*
6. sine re *pro* res.   super *pro* numero.
7. Dic ergo 10 res.   coequant.   Sume
medietatem radicum.
8. in semetipsam (semetipsa D).
9. A quibus 21 demptis V; *om.* D.   et *om.* V.
Accipe eorum (earum D) radicem, id est 2, et
eam ex.
10. et *om.* D.
13. ut (et D) unaquaque diuisione in semetipsa
(semetipsam D) deducta, si multiplicatio minoris
diuisionis ex multiplicatione maioris tollatur 40
(*om.* D) remaneant.
16. Expositio huius talis.   consimili.
17. 100 et substantia absque (+ absque vel

minus *in marg.* D *man.* 1) 20 radicibus. Multiplica igitur rem in re, et fiet.
18. que V.   absque *om.* V; et D.   20
rebus.
20. 100 cum 20, et eas super 40 adicias.
21. drachmas *om.* Hoc ergo centeno opponas
(apponas D) numero et 40 ex. inferas D.
22. 60 (4, + cum D) 20 res coequancia.
23. unam mensurat diuisionem.
25. unaquaque.   semetipsa.   multiplica
D.   *In marg.* $\lfloor 3 + \phi \; 7\!\!\!Q$ D.
26. et utriusque multiplicatione in unam col-
lecta (+ a D) quantitate que.
27. addita + nam et unamquamque earum
in semetipsa multiplicasti D.
28. Inde igitur talis datur expositio, ut.   consimili.

First Problem[1]

I divide ten into two parts in such a way that the product of one part multiplied by the other gives 21.[2]

Now then we let x represent one part, which we multiply by $10 - x$, representing the other part. The product $10x - x^2$ is equal to 21 units. Complete $10x$ by x^2 and add this x^2 to 21. This gives $10x$ equal to $x^2 + 21$ units. Take one-half of the unknowns, *i.e.* 5, and multiply this by itself, giving 25. From this subtract 21, giving 4. Take the root of this, 2, and subtract it from half of the roots, leaving 3, which represents one of the parts.

Second Problem

I divide ten into two parts in such a way that each part being multiplied by itself, the product of the smaller part taken from the product of the larger part leaves 40.[3]

Explanation. You multiply $10 - x$ by itself, giving $100 + x^2 - 20x$. You multiply x by itself, giving x^2, which you take from $100 + x^2 - 20x$, leaving $100 - 20x$ equal to 40 units. Therefore by adding $20x$ to the 40 units complete 100 units by $20x$. This gives 100 equal to 40 units $+ 20x$. Therefore you take 40 from 100, leaving 60 units equal to $20x$. Three is then the value of x and represents one part.

Third Problem

I divide ten into two parts in such a way that when to the sum of the products of each part by itself is added the difference between the two parts the sum total will be 54 units.[4]

---

[1] The sixteen problems which follow are selected by Robert of Chester from twice that number in the Arabic text. The Boncompagni version presents nine, including the first, second, third, and fifth of this list.

The Boncompagni version interjects (*loc. cit.*, pp. 45–46) before these problems: "pro multitudine data assignatur horum muttatione quaslibet questiones secundum restauracionem propositas in predictos modos solubiles esse palam est. Cuius utilitas ad documentum libri elementorum precipua est, in inveniendis scilicet lineis alogis et medialibus binomiis et residuis sive reccisis que per notum numerum assignari non possunt. In practica quoque geometrie et universis questionibus ignotorum secundum arismeticam formatis certissima via est."

[2] $x(10 - x) = 21$, which leads to the type equation given earlier in the work, $x^2 + 21 = 10x$, and indeed, it incidentally appears a second time in the preceding set of problems. The Arabic text precedes the statement of this problem, Rosen, *op. cit.*, p. 41, as follows: "If a person puts such a question to you as . . ."; the Boncompagni version, *loc. cit.*, p. 46, precedes: "Igitur sub formas precedencium et alias questiones proponimus. Queritur. . . ."

[3] $(10 - x)^2 - x^2 = 40$, whence $100 - 20x = 40$; $20x = 60$, $x = 3$, which is the value of the root.

[4] $x^2 + (10 - x)^2 + 10 - 2x = 54$, whence $2x^2 - 22x + 110 = 54$; $2x^2 + 110 = 54 + 22x$; $x^2 + 55 = 27 + 11x$; $x^2 + 28 = 11x$; $\frac{1}{2}$ of 11 is $5\frac{1}{2}$, $(5\frac{1}{2})^2 = 30\frac{1}{4}$, $30\frac{1}{4} - 28 = 2\frac{1}{4}$, $\sqrt{2\frac{1}{4}} = 1\frac{1}{2}$, $5\frac{1}{2} - 1\frac{1}{2} = 4$, which is the value of the unknown quantity.

1 producentur 100 ex numero et vna substantia, absque 20 rebus. Rem etiam cum
re multiplices, et producetur substantia, quam 100 ex numero et substantiae,
absque 20 rebus iunge, et venient 100 ex numero et 2 substantiae, absque 20
rebus. Iam quantitatem quae inter vtramque partem est, 10 sine 2 rebus, toti
5 summae adde, atque hoc totum ad 110 ex numero et 2 substantias, duabus et
20 rebus abiectis, 54 drachmas coaequantes, exuberando peruenit. Compleas ergo,
et dic: 110 drachmae et 2 substantiae 54 drachmas et 22 res coaequant. Hoc
autem ad vnam conuertas substantiam, et dic: 55 drachmae et vna substantia
coaequant 27 drachmas et 11 res. Igitur 27 ex 55 subtrahe, et manebunt 28 drachmae
10 et vna substantia, 11 res coaequantia. Res igitur per medium diuide, et venient
5 et medium; haec cum suo aequali multiplica, et producentur 30 et quarta, ex
his 28 subtrahe, et radicem ex residuo, vnum scilicet et $\frac{1}{2}$ accipe, quae simul ex
medietate rei auferas, et manebunt 4. Haec igitur vnam partem diuisionis adimplent.

15
Quaestio quarta

Denarium numerum sic in duo diuido, vt vna eius parte cum seipsa multiplicata,
numerus inde productus alteram octagesies et semel comprehendat.

Expositio est, vt 10 sine re cum suo aequali multiplices, et producentur 100 ex
numeris et substantia, absque 20 rebus, 81 res coaequantia. Comple igitur 100
20 ex numeris et substantiam, et adde 20 res numero 81, et venient 100 ex numeris
et substantia, 100 res et vnam coaequantia. Mediabis ergo res, et venient 50
et dimidium. Haec cum suo aequali multiplica, et producentur 2550 et quarta.
Ex his nunc 100 subtrahe, et manebunt 2450 et quarta. Horum radicem accipe,
id est 49 et dimidium, atque hanc ex radicum medietate subtrahe, et manebit
25 vnitas, vnam partem numeri decem ostendens.

Quaestio quinta

Duas substantias duabus drachmis differentes, diuido, maiorem scilicet in

1. erit.   vna *om.*
1–4. Rem . . . rebus [1] *om.* V; Rem quoque in
re multiplicata, et erit substantia. Hec insimul
iunge et erunt c et due substantie absque xx
rebus D.
4. Et iam.   vtramque + substantiam D.
super totam summam addidisti. Dicas ergo
quantitas que inter utramque est diuisionem 10
absque duabus rebus designat. Hoc igitur totum
ad 100 et 10 et (*om.* D) duas *pro* 10 sine . . . et 2.
5. absque duabus rebus in (et D) *pro* duabus et.
6. Tunc ergo compleas.
8. conuerte.   vna *om. et sic infra.*
10. erunt.
11. haec + igitur.   consimili.   erunt.
12. his + ergo.   subtrahas, et radicem (radices D) remanentis (remanentes D) id est duorum
et 4e accipias (+ Accipe radicem D) id est unum et
medium que simul.
13. radicis V; radices D *pro* rei.   igitur + 4.

16. et *pro* vt D.   in se ipsam semel deducta
(deducto D), 81 vicibus alteri (iubet altera D) com-
prehendetur.
18. Expositio + huius rei.   duo *pro* suo D.
consimili.   ex numeris *om. saepe.*
19. minus.
20. et substantiam *om.*   et [2] *om.* D.   radices super (*om.* D). 81, et erunt.   100 *om.* D.
21. et substantia *om.* D.   100 et unam radicem.   radices.   erunt.
22. Has ergo.   suo consimili, sibi aequali =
cum seipsis: se *in marg.* C; in suo consimili V D.
erunt 2000, 500 et 50 et 4ª.
23. ergo *pro* nunc.   2000 et 450 (400 D).
Horum + ergo.
24. et ipsam *pro* atque hanc.   medietate
+ id est 50 et medio.   subtrahas.
25. unum, unam denarii ostendens diuisionem.
27. diuido *om.* V.   id est earum minorem
super maiorem.

The explanation of this problem is of the following nature : you multiply $10 - x$ by its equal, which gives $100 + x^2 - 20x$. Also you multiply x by x, giving x^2, which you add to $100 + x^2 - 20x$. This gives $100 + 2x^2 - 20x$. To the sum total now add the difference between the two parts, $10 - 2x$, and this total amounts to $110 + 2x^2 - 22x$, which equals 54 units. You complete therefore (by adding $22x$), and you obtain 110 units $+ 2x^2$ equal to 54 units $+ 22x$. This, moreover, you reduce to one square, giving 55 units $+ x^2$ equal to 27 units $+ 11x$. So subtract 27 from 55, giving 28 units $+ x^2$ equal to $11x$. Halve the unknowns, giving $5\frac{1}{2}$. Multiply this by itself, giving $30\frac{1}{4}$. From this subtract 28 and the root of the remainder, $1\frac{1}{2}$, taken from one-half of the roots, will leave 4. Therefore this represents one part.

Fourth Problem [1]

Divide ten into two parts in such a way that the product of one part by itself equals 81 times the other part.[2]

Explanation : you multiply $10 - x$ by itself, giving $100 + x^2 - 20x$, which equals $81x$. Complete $100 + x^2$ by adding $20x$ to $81x$. This gives $100 + x^2$ equal to $101x$. You halve the unknowns, giving $50\frac{1}{2}$. Multiply this by itself, giving $2550\frac{1}{4}$. Now subtract 100 from it, leaving $2450\frac{1}{4}$. Take the root of this, i.e. $49\frac{1}{2}$, and subtract it from the half of the roots. This will give unity, representing one part of ten.[3]

Fifth Problem

Two squares (i.e. two quantities or numbers) being given with a difference of two units, I divide the smaller by the larger in such a way that the fraction resulting from the division shall equal one-half.[4]

[1] The order of the problems is from this point different from the order of the corresponding set of problems as given in the Libri, Arabic, and Boncompagni versions, but the variations are due to the omission of problems, as will be noted below, rather than to actual changes in order. See chapter V of our Introduction. The Boncompagni text is the least complete, giving only ten problems, which follow, in the main, those of the Libri text.

[2] This problem $(10 - x)^2 = 81x$, appears on pages 47–48 of Rosen's book, and appears twice in the Libri text, pp. 282–283 and pp. 289–290.

In modern notation, the solution proceeds :

$100 - 20x + x^2 = 81x$; $x^2 + 100 = 101x$; $\frac{1}{2}$ of 101 is $50\frac{1}{2}$, $(50\frac{1}{2})^2 = 2550\frac{1}{4}$; $2550\frac{1}{4} - 100 = 2440\frac{1}{4}$; $\sqrt{2440\frac{1}{4}} = 49\frac{1}{2}$; $50\frac{1}{2} - 49\frac{1}{2} = 1$, which is the value of one root. The other root would be $50\frac{1}{2} + 49\frac{1}{2}$, or 100.

[3] The second root, 100, is not given, since it leads to 100 and negative 90 as the two parts into which 10 is divided. This was rejected by the Arabic writer as an impossible solution, nor, indeed, was such a solution regarded as possible for centuries after the time of the Arab.

[4] Rosen, pp. 50–51, and again pp. 62–63; Libri, p. 283 and p. 295. By error Scheybl makes this one-half of the larger square.

1 minorem, sic, vt vna diuisionis particula, quae est exiens, medietatem substantiae maioris compleat.

Rem ergo et duas drachmas cum medietate quam diuisionis particulam diximus, multiplico, et veniet rei medietas et drachma, rem coaequans. Propter medie-
5 tatem ergo rei vtrinque medietatem rei abiicio, et manebit drachma medietatem rei coaequans. Hanc igitur duplico, et venient duae drachmae, quae ipsam rem, constituunt, vnde quatuor deinde alteram.

Quaestio sexta

Substantiam et eius radicem sic multiplico, vt multiplicationis productum tres
10 similitudines substantiae, adimpleat.

Expositio est, quoniam quando radicem cum tertia substantiae multiplicaueris, tota producitur substantia. Tria igitur huius substantiae radicem adimplent, et 9 ipsam substantiam.

Quaestio septima

15 Tres radices substantiae cum quatuor eius radicibus ita multiplico, vt tota multiplicationis summa ipsi substantiae et quadraginta quatuor drachmis coaequetur.

Expositio talis est, vt cum quatuor radicibus substantiae tres eius radices multiplices, et producentur 12 substantiae, substantiam vnam et 44 drachmas
20 coaequantes. Minue igitur substantiam ex 12 manebunt 11 substantiae, 44 drachmas coaequantes. Substantia ergo vna 4 aequat.

Quaestio octaua

Quatuor substantiae radices cum quinque eiusdem substantiae radicibus sic multiplico, vt tota multiplicationis summa ipsius substantiae duplo et triginta
25 sex drachmis coaequetur.

Quatuor radices substantiae cum suis quinque radicibus multiplico, produco autem sic 20 substantias, duas substantias et 36 drachmas coaequantes. Ex 20 igitur substantiis duas substantias aufero, et manebunt 18 substantiae 36 drachmas coaequantes. Igitur 36 in 18 diuido, et exeunt 2, quae ipsam substantiam adimplent.

1. sic diuido, ut unam diuisionis particulam compleat medietas.

4. fiet.   coequantes D.   Cum medietate; C corr. in P C.

5. vtrinque om.   rei² om. V.   proicio. drachma om. V.

6. Hoc.   dico pro venient.   quae om.; in marg. C.   rem + hoc est vnam substantiam vel numerum vnum C.

7. et alteram 4.

9. in pro et.

10. substantiae + hoc tres substantias propositas C.

11. Expositio + huius.   substantiam ante substantiae D.

12. totam D.   oriatur.   igitur om.

13. ipsam om.

15. et pro vt C.

16. ipsa D.

18. Huius est expositio, ut 4 eius radices in suis tribus radicibus.

20. 12 + et.   substantiae + et V.

21. coequans V.   Substantiam V; Substantie D.   coequant.

23. eius.   substantiae om.

24. vt + et.   ipsi V; ipse D.   duplo om.

25. xxxvi radicibus D.

26. Expositio est, vt quatuor C.   Quatuor + ergo.   substantiae om.   multiplica V.   et erunt 20 substantie.

28. xi pro 18 et xliiii pro 36 D.

29. coaequantes + ex xx ergo substantiis duas substantias aufero et remanebunt xviii substantie D.   36 super (sunt D) 18.   erunt.   substantiam om. D.

I multiply $x + 2$ units by $\frac{1}{2}$, representing the quotient.[1] This gives $\frac{1}{2} x$ and a unit equal to x. On account of the one-half x I take one-half x from both sides, leaving a unit equal to $\frac{1}{2} x$. I double this, giving two units, which equal the unknown. Whence four is the other.

Sixth Problem

I multiply a square by its root in such a way that the product equals three similar squares.[2]

Explanation. From what is given it follows that when you multiply the root by one-third of the square, the total gives the square.[3] Therefore 3 represents the root of this square, and 9 the square.

Seventh Problem

I multiply three roots of a square by four of its roots in such a way that the sum total of the multiplication will equal the same square and 44 units.[4]

Explanation. Multiplying four roots of a square by three of its roots, we have twelve squares, which are equal to one square and 44 units.[5] Take therefore the one square from the 12 squares, leaving 11 squares equal to 44. Hence one square equals 4.

Eighth Problem

I multiply four roots of a square by five roots of the same square in such a way that the sum total of the multiplication will equal double the square and 36 units.[6]

Explanation. I multiply four roots of a square by five of its roots and I have as a product 20 squares, which are equal to 2 squares and 36 units. Therefore I take the two squares from the 20 squares, leaving 18 squares equal to 36 units. I divide 36 by 18, and obtain 2, which represents the square.[7]

---

[1] $\dfrac{x}{x + 2} = \frac{1}{2}$, whence $x = \frac{1}{2} (x + 2)$; $x = \frac{1}{2} x + 1$; $\frac{1}{2} x = 1$, $x = 2$.

[2] Rosen, pp. 54–55; Libri, p. 291.

Al-Khowarizmi carefully avoids the term for the cube of the unknown, with which he was certainly familiar, for Diophantus and other Greek writers employed the term. His continuator, Abu Kamil, discussed not only cubics which, like this one, are reducible immediately (to equations of lower degree), but also equations in quadratic form. However, systematic discussion of the general cubic was first attempted by Omar Khayyam.

[3] $x^2 \cdot x = 3 x^2$.

[4] Rosen, p. 55; Libri, p. 291.

[5] $3 x \cdot 4 x = x^2 + 44$.

[6] Rosen, p. 55; Libri, p. 284.

[7] $4 x \cdot 5 x = 2 x^2 + 36$; $18 x^2 = 36$; $x^2 = 2$.

¹

Quaestio nona

Radicem substantiae cum eius quatuor radicibus ita multiplico, vt tota multiplicationis summa tribus substantiis et quinquaginta drachmis coaequetur.

Expositio est, vt radicem cum quatuor radicibus multiplicem, et 4 substantias
5 ita productas tribus substantiis et 50 drachmis coaequem. Ex 4 igitur substantiis tres substantias aufero, et manebit substantia, 50 drachmas coaequans. Igitur radix substantiae 50 est radix numeri 50.

Quaestio decima

Ex substantia eius partem tertiam et tres drachmas aufero, et quod residuum
10 fuerit cum suo consimili multiplico, et oritur ex multiplicatione ipsa substantia.

Expositio est, vt quando tertiam et tres drachmas subtraxero, duae tertiae absque tribus drachmis maneant. Sunt autem res. Vnde duas tertias rei absque
3 drachmis cum suo aequali multiplicamus, et producuntur quatuor nonae substantiae et 9 drachmae, absque quatuor radicibus, radicem coaequantia. Adde
15 igitur absque 4 radicibus vni radici, et venient quatuor nouenae substantiae et 9 drachmae quinque radices coaequantes. Oportet ergo vt quatuor nouenas substantiae compleas, vt ipsa substantia perficiatur, et hoc quidem vt singula cum 2 et quarta multiplices, et venient 11 res et $\frac{1}{4}$, vnam substantiam et 20 drachmas cum quarta coaequantes. Operare igitur cum eis quemadmodum in medietatione
20 radicis tibi diximus.

Quaestio vndecima

Tertiam substantiae cum eius quarta sic multiplico, vt tota multiplicationis summa ipsi substantiae coaequetur.

Expositio huius est, vt tertiam rei cum quarta multiplicem, et producetur
25 medietas sextae vnius substantiae, rem coaequans. Radix igitur substantiae in 12 terminatur, et substantia 144 in se continet.

3. tribus + similitudinibus.   substantie.
coequatur V; adequatur D.

4. Huius est expositio.   et erunt 4 substantie, 3 substantias.

5. coequantes; C add. in marg. aequales eodem preferam.   4 + ex D.   igitur om.

6. tollo.   coaequans + et ipsa et radix l' in iiii^{or} radicibus l' D.   Radix ergo 50 in 4 radicibus 50 ducta (sunt D) ducenta que tribus similitudinibus substantie et 50 dragmatibus equantur constituit (manet equalis D).

7. C add.: Sequitur examen.

Radix $\sqrt{50}$.   Substantia 50.
$\sqrt{50}$ vt vna radix, cum $\sqrt{800}$, quatuor radicibus, multiplicata, producit 200. Atque tantum sunt etiam 50 ter, hoc est tres substantiae, et quinquaginta.

9. partem om.   tollo.

10. et om.   ex + qua videlicet.

11. Huius est (om. D) expositio; Huius C sed del.

12. remanebunt, et ipse erunt radix. Unde (+ et V).

13. multiplico (multiplicamus D) + Due ergo 3^e in duas 3as ducte 4 substantie nouenas componunt et cum (+ et cum D) tribus in duabus rei 3^{iis}multiplicate, duas res componunt similiter (+ duas D) diminutiuas et sine 3^{bus} in 3^{bus} 9 dragmata constituunt adiectiua. Erunt ergo. nouene.

14. 8 pro 9.

15. sine 4 radicibus super radicem, et fient. nouenas.

16. coequans.

17. vt¹ + et.   compleatur et (om. D) hoc est ut illas in duabus (duobus D) et (aut D) 4ta.

18. multiplices (+ vel, si placet, singula in 4/9 diuidas C). Multiplica ergo 9 in duo et 4^a et erunt xx (11 res V) et 4^a (+ et multiplica v radices in duo et 4ta et erunt 11 res et 4ta D).   Habebis ergo substantiam et 20 dragmata et quartam, 11 res et 4^{am} rei coequancia. Oppone ergo.

20. radicum.

24. tertiam + et D.   deducam V; deductam D.

25. sextae + hoc est vna duodecima C. rei add. super substantiae et numerum super rem C.

Ninth Problem

Multiply the root of a square by four of its roots in such a way that the sum total of the multiplication shall equal three squares and 50 units.[1]

Explanation.   I multiply the root by 4 roots and I obtain four squares which are equal to three squares and 50 units.   I take the 3 squares from the 4 squares, leaving a square equal to 50 units.[2]   Therefore the root of this square is the root of 50.[3]   The root of 50 multiplied by 4 roots of 50 gives 200, which is equal to three of the squares and 50 units.

Tenth Problem

I take from a square one-third of it and three units, and multiply the remainder by itself; the product is the square.[4]

Explanation.   When I have subtracted one-third and three units, two-thirds less three units remain.   Now let the square be represented by x; then we multiply $\frac{2}{3} x - 3$ units by itself and have $\frac{4}{9} x^2 + 9$ units $- 4$ roots equal to one root.   Add the one root to the four roots, giving $\frac{4}{9} x^2 + 9$ units equal to 5 roots.   It is now necessary to complete the four-ninths of a square, so as to make it a whole square.   You multiply each side by $2\frac{1}{4}$,[5] giving $11\frac{1}{4} x$ equal to $x^2 + 20\frac{1}{4}$ units.   You perform the operations with these, then, in the manner which we have explained to you in the sections on the halving of the root.

Eleventh Problem

I multiply one-third of a square by one-fourth of it in such a way that the sum total of the multiplication will give the square.[6]

Explanation.   I multiply $\frac{1}{3} x$ by $\frac{1}{4} (x)$ and I obtain $\frac{1}{2}$ of $\frac{1}{6} x^2$ equal to x. The root of the square, then, is 12, and the square is 144.[7]

---

[1] Rosen, p. 56; Libri, pp. 291–292.

[2] $x \cdot 4 x = 3 x^2 + 50$.

[3] Scheybl adds: "The root of fifty multiplied by four roots of 50 gives 200, which is equal to three of the squares and 50 units."   Robert of Chester's text is more closely followed by the Vienna and Dresden manuscripts.

The symbol for square root $\sqrt{}$ used by Scheybl was introduced by Adam Riese in his *Coss* written in 1524, but not printed until recently (Berlet, Leipzig, 1892); for the word *Coss* see page 38.

[4] Rosen, pp. 56–57; Libri, pp. 284–285.   In modern notation: $(x - \frac{1}{3} x - 3)^2 = x$.   Between this and the preceding problem the Arabic text includes a problem, leading to the equation:

$$x^2 + 20 = 12 x.$$

[5] Scheybl adds that you may, if you prefer, divide both sides by $\frac{4}{9}$.

[6] Rosen, p. 58; Libri, p. 292. $\frac{1}{3} x \cdot \frac{1}{4} x = x$.

[7] In the Arabic text this problem follows: $(\frac{1}{3} x + 1)(\frac{1}{4} x + 2) = x + 13$.

1

Quaestio duodecima

Drachmam et medium ita in duo diuido, vt maior pars duplex minori habeatur. Expositio huius est, vt maiorem partem ad minorem vnum et rem constituam; vnde etiam dicam drachmam et dimidium super drachmam et rem diuisa, et duae res 5 drachmam constituunt. Duas igitur res cum drachma et re multiplico, producuntur autem duae substantiae et duae res, drachmam et dimidium coaequantes. Eas igitur ad vnam conuerto substantiam, hoc est, vt ex omni re suam auferam medietatem. Dico ergo, substantia et res tres quartas drachmae coaequant. Operare nunc cum his quemadmodum tibi iam diximus.

10

Quaestio decima tertia

Substantiam cum eius duabus tertiis multiplico, et fiunt quinque.

Expositio talis est, vt rem cum duabus tertiis rei multiplicem, et producuntur duae tertiae substantiae quinque coaequantia. Comple igitur $\frac{2}{3}$ substantiae cum similitudine earum medietati, et veniet substantia. Similiter comple 5 cum sua 15 medietate, et habebis $7\frac{1}{2}$ quae substantiam coaequant. Ipsam igitur substantiae radicem quae cum suis duabus tertiis multiplicata quinarium numerum producit.

Quaestio decima quarta

Inter puellas drachmam sic diuido, vt vnicuique earum aequalis particula rei contingat; quibus etiam si vnam insuper puellam adhibuero, illis omnibus par-20 ticula primae particulae minus vna sexta aequalis contingit.

Expositio huius est, vt ipsas puellas cum minutia particulae qua differunt multiplicem, postea quod ex multiplicatione excreuerit cum numero puellarum postremarum multiplicem atque tandem productum hoc in id quo puellae priores a posterioribus differunt, diuidam, et complebitur substantia. Puellas igitur 25 priores, vnam rem scilicet, cum sexta, quae est inter eas, multiplico, et producitur $\frac{1}{6}$ rei. Deinde multiplico eam cum numero puellarum postremarum, quae sunt res et vnum, et producitur sexta substantiae et sexta rei in drachmam diuisa, drachmae aequalis. Substantiam igitur compleo, id est, substantiam cum senario numero multiplico, et sic substantiam et radicem habebo. Drachmam etiam cum 30 sex drachmis multiplico, et venient substantia et radix, sex drachmas coaequantes.

2. per 2 V; in duo inequaliter D.

3. constituas V.

4. et *pro* etiam.   diuisi V; diui[s]um D.

5. et re *om.* V.   et (*om.* V) fiuntque.

6. autem *om.*

7. conuerte.   et hoc.

8. Dicam.   tres *om.*

9. Oppone ergo.   hiis D.

11. sic multiplico, ut fiant.

12. talis *om.*

13. unius substantie 5.

14. erit.   Et similiter.   unum *pro* v D *del.*

15. substantia *pro* medietate D.   quae *om.* substantias D.   coequantem.   Eius ergo radix est res que quando in.

16. fuerit ad quinarium excrescet numerum. C *add.*: radix numeri quae est $\sqrt{7\frac{1}{2}}$ componet.

18. unum dragma.   uniuscuiusque.

19. insuper *om.*   His *pro* illis D.

20. minus sexta tocius.   contingat.

21. ut primas puellas V; *om.* D.   cum *om.* V.   que D.

22. Ac postea.   numerum V.

23. posteriorum D.   multiplicabo (multitudo D). Postea quod ex hac multiplicatione collectum fuerit super illud quod (+ quod V) inter puellas primas est (*om.* D) et posteriores diuido.

24. diuidunt C.   et + tunc.   Tunc etiam puellas primas que sunt res in sexta.

25. multiplica.

26. radicis.

27. 5 *pro* vnum.   radicis.

28. coequalis.   ipsam *pro* substantiam.[2]

29. ergo in 7 *pro* etiam cum sex.

30. erunt.

Twelfth Problem

I divide a unit and one-half in such a way that the larger part shall be double the smaller.[1]

Explanation. I let the greater part be to the lesser as one is to x; whence I say divide $1\frac{1}{2}$ units by one unit $+ x$, giving $2x$. Therefore I multiply $2x$ by one unit $+ x$, giving $2x^2 + 2x$, which is equal to $1\frac{1}{2}$ units. I reduce this, therefore, to one square, that is, of each thing I take the half. I say then that $x^2 + x$ is equal to $\frac{3}{4}$ of a unit. You now proceed in the manner which we have explained.

Thirteenth Problem [2]

I multiply a square by two-thirds of itself and have five as a product.[3]

Explanation. I multiply x by two-thirds x, giving $\frac{2}{3}x^2$, which equals five. Complete $\frac{2}{3}x^2$ by adding to it one-half of itself, and one x^2 is obtained. Likewise add to five one-half of itself, and you have $7\frac{1}{2}$, which equals x^2. The root of this, then, is the number which when multiplied by two-thirds of itself gives five.

Fourteenth Problem

I divide a unit among girls in such a way that each one receives the same fractional part of the thing. Now if I add one girl to the number, each receives for her part one-sixth (of a unit) less than before.[4] .

Explanation. I multiply the number of girls at the first by the fractional part representing the difference. Then I multiply this product by the second number of girls, and finally I divide this product by the difference between the first and second number of girls. This completes the given quantity.[5] Hence I multiply in this instance one x, representing the first number of girls, by the difference between the two amounts, $\frac{1}{6}$, and $\frac{1}{6}x$ is obtained. Then I multiply this product by the second number of girls, which is $1 + x$, and $\frac{1}{6}x^2 + \frac{1}{6}x$ is obtained; this being divided by a unit equals a unit. I complete the square, that is, I multiply the square by six, and I have $x^2 + x$. Likewise I multiply the unit by six units, giving $x^2 + x$

---

[1] Strictly, $1\frac{1}{2} - x = 2x$, whence $x = \frac{1}{2}$. The English translation of this problem is adapted by Rosen (p. 59) to conform to the solution in the explanation, and this follows closely our explanatory text. The equation which is given by Rosen is: $\dfrac{1\frac{1}{2}}{1+x} = 2x$, and to this our text leads.

I am indebted to Professor W. H. Worrell for the following precise translation of the Arabic text of the passage: "If it is said to divide a unit and a half between a man and a part of a man, then the man has received the like of the fraction." The Libri text, pp. 285–286, varies from this only in stating that the man receives double that which the fractional part (of a man) receives.

[2] The following problem precedes in the Arabic text: $(x - \frac{1}{3}x - \frac{1}{4}x - 4)^2 = x + 12$.

[3] $x \cdot \frac{2}{3}x = 5$, whence $x = \sqrt{15/2}$.

[4] Rosen, pp. 63–64; Libri, p. 286. Our text is faulty. The problem is $\dfrac{1}{x} - \dfrac{1}{x+1} = \frac{1}{6}$.

[5] Latin *substantia*, 'square,' obviously an error.

1 Radices igitur per medium diuido, et earum medium cum suo aequali multiplico, et quod producitur ad 6 adiicio, atque huius aggregati radicem accipio, vnde tandem medietatem radicum subtraho. Nam hoc quod residuum fuerit, numerum puellarum priorum designabit, et ipsae sunt duae.

5 *Quaestio decima quinta*

Si ex substantia quatuor radices subtraxero, ac postea tertiam residui accepero, et haec ipsa tertia quatuor radicibus aequalis fuerit, tunc substantia in ducenta quinquaginta sex terminatur.

Expositio huius est vt scias, quod cum tertia residui, posterioris scilicet, aequalis 10 sit quatuor radicibus, residuum prius duodecim radicibus aequale erit. Adde igitur illas super quatuor, et venient 16 radices, quae sunt radix substantiae.

Quaestio decima sexta

Ex substantia tres radices subtraho et postea quod residuum est cum suo aequali multiplico, sitque tota multiplicationis summa aequalis substantiae.

15 Manifestum est, quòd residuum sit radix, quam quaternarius adimplebit numerus; substantiam verò numerus 16 component.

Hae igitur sunt sedecim quaestiones quae ex prioribus nasci videntur, vt diximus. Quicquid igitur iuxta artem restaurationis et oppositionis multiplicare volueris, facile per ea quae tradita sunt expedies.

20 **CAPUT RERUM VENALIUM**

Res autem venales et omnia, quae ad ipsas attinent, duobus modis et quatuor numeris disponuntur. Horum igitur numerorum primus iuxta Arabes, Almuzahar, qui et primus propositus nominatur. Secundus vero, Alszian, id est secundus per primum dinotus, appellatur. Tertius Almuhen id est ignotus. Quartus Alche-25 mon, id est per primum et secundum dinotus. Hi porro quatuor numeri sic disponuntur, vt eorum primus, qui est Almuzahar vltimo, qui est Alchemon, opponatur. Horum autem quatuor numerorum tres semper noti ac certi ponuntur, quatuor vero numerus ignotus ponitur, et is ipse est cum quo quantum inquiritur.

2. atque eam multiplicationem super 6 adicio, et huius summe.   ex qua *pro* vnde tandem.

3. abstraho.   Nam + et.

4. et sunt due V; *om.* D.

6. ex *om.* D.   et postea; et C *in ras.*   tertiam *om.* D.

7. si *pro* et haec.   et *pro* in D.

9. quoniam *pro* quod cum.   posterioris scilicet *om.*

10. et quam residuum simile sit (est D) 12 radicibus.

11. quatuor + radices.

13. ac *pro* et.   est + et D.

15. Unde manifestum est, quod illud residuum similiter sit radix et quod substantia sit 4 et fiunt (fuerit D) 16 dragmata.

17. Hec.   ex 6 primis.   vt + iam.

18. Quotquot V; Quidquid D.

19. per earum aliquam illud multiplicatum reperies.

20. *Titulum om.*

21. Item res omnis venales D.   ad *om* ipsis.

22. vero *pro* igitur.   Almuzarar *siue* Almusarar *ubique* V; Almuzaar *siue* Almusaar D.

23. id est *pro* qui et.   nuncupatur D. Alter Alszarar *ubique* V; Alzazar *siue* Alszazar D.

24. Almuthemen *ubique* V; Almute *siue* Almuthemon D.   Althemen *ubique* V; Altemon *siue* Althemon D.

25. Sed et hii 4.

27. eciam.

28. numerus *om.*   ponitur + et   incertus. ipse est ille.

equal to 6 units. I take one-half of the roots and I multiply the half by itself. I add the product to 6, and of this sum I take the root. The remainder obtained after subtracting one-half of the roots will designate the first number of girls, and this is two.

Fifteenth Problem

If from a square I subtract four of its roots and then take one-third of the remainder, finding this equal to four of the roots, the square will be 256.[1]

Explanation. Since one-third of the remainder is equal to four roots, you know that the remainder itself will equal 12 roots. Therefore add this to the four, giving 16 roots. This (16) is the root of the square.

Sixteenth Problem

From a square I subtract three of its roots and multiply the remainder by itself; the sum total of this multiplication equals the square.[2]

Explanation. It is evident that the remainder is equal to the root, which amounts to four. The square is 16.

These now are the sixteen problems which are seen to arise from the former ones, as we have explained. Hence by means of those things which have been set forth you will easily carry through any multiplication that you may wish to attempt in accordance with the art of restoration and opposition.

CHAPTER ON MERCANTILE TRANSACTIONS[3]

Mercantile transactions and all things pertaining thereto involve two ideas and four numbers.[4] Of these numbers the first is called by the Arabs Almuzahar and is the first one proposed. The second is called Alszian, and recognized as second by means of the first. The third, Almuhen, is unknown. The fourth, Alchemon, is obtained by means of the first and second. Further, these four numbers are so related that the first of them, the measure, is inversely proportional to the last, which is cost. Moreover, three of these numbers are always given or known and the fourth is unknown, and this

---

[1] Rosen, p. 66; Libri, p. 296. $\frac{1}{3}(x^2 - 4x) = 4x$.

In the Arabic text these two problems precede: $x^2 . 3x = 5x^2$ and $(x^2 - \frac{1}{3}x^2) . 3x = x^2$.

[2] Rosen, p. 67; Libri, p. 296. $(x^2 - 3x)^2 = x^2$, whence $x^2 - 3x = x$.

The problem, $x + \sqrt{x^2 - x} = 2$, precedes. This is one of two problems given in the German excerpt of 1461 from the algebra of Al-Khowarizmi (Gerhardt, *Monatsbericht d. Königl. Akad. der Wissenschaften zu Berlin*, 1870, pp. 142–143).

[3] The famous 'rule of three' is the subject of discussion in this chapter.

[4] The two ideas appear to be the notions of quantity and cost; the four numbers represent unit of measure and price per unit, quantity desired and cost of the same. These four technical terms are *al-musa' 'ir, al-si'r, al-thaman,* and *al-muthammin,* and further *al-maqūl;* see p. 44.

1 Talis igitur ad hanc artem regula datur, vt in omni huius inquisitione tres numeri, qui noti ac certi positi sunt, considerantur, quoniam eorum duo semper inter se oppositi inueniuntur. Horum igitur duorum vnus cum altero multiplicandus, atque multiplicationis productum in notum tertium ac certum positum, qui 5 ignoto opponitur, diuidendum erit. Nam quod ex diuisione exierit, erit numerus de quo dubitatur, et ipse ei numero opponitur in quem facta est diuisio. Sed ne hanc artem aliquis error incurrere arbitretur, tale damus exemplum.

De primo modo

Decem pro sex, quot pro quatuor?

10 Vide nunc, quo modo, pro eo vt diximus, praefati numeri disponuntur. Nam quando 10 dixisti, numerum Almuzahar pronunciasti; et quando pro 6 dixisti, Alszian protulisti; et quando quot dixisti, numerum Almuhen siue Magol, id est ignotum, pronunciasti; et quando pro 4 dixisti, numerum Alchemon edidisti. Vides igitur quòd eorum tres, id est 10, 6 et 4, noti et certi sint, de quarto verò 15 adhuc ignoto, dubitetur. Si igitur ad regulam prius datam respexeris, primum cum vltimo, id est 10 cum 4, multiplicabis, sunt etenim oppositi numeri, noti quoque ac certi, et quod ex multiplicatione excreuerit, id est 40, in alterum numerum notum ac certum, qui est Alszarar, id est in 6 oportet diuidere, et exeunt 6 et $\frac{2}{3}$ vnius, numerum ignotum designantes. Et hic numerus numero senario, qui 20 Arabice Alszarar nominatur, est oppositus.

De secundo modo

Secundus modus huius artis est, vt dicas, decem pro 8, pro quot quatuor?

Decem igitur sunt Almuzahar, qui videlicet numero Almuhen ignoto, cum quo quantum acquiritur, est oppositus, et 8 designat numerum Alszarar, qui numero 25 Alchemon, qui sunt 4 opponitur. Vnum igitur duorum numerorum cognitorum atque oppositorum cum altero multiplica, id est 4 cum 8, et producentur 32. Haec ergo 32 in tertium cognitum numerum 10, qui est Almuzahar, diuide, et exeunt $3\frac{1}{5}$, qui numerum Alchemon designant, quique ei numero in quem diuiditur est oppositus.

1. quoque *pro* igitur.   hac D.

2. certe V.   quoque *pro* quoniam D.   adinuicem *pro* inter se.

3. inuenientur.   unus est; unius C.

4. et eorum *pro* atque.   per.   atque V. certe D.

5. diuidenda V; diuidendus D.

6. dubitabatur.   per.

7. errorem.

8. *Titulum om. et sic infra.*

9. Secundum ergo primum modum, sic dicas, 10 pro 6, quot pro 4?

10. secundum quod diximus.

11. quoniam *pro* quando.

12. Alszaran C.   Magul V; Magulum D.

13. nunciasti.   Almuhemen C.

14. qualiter *pro* quod.   10 + et D.   ac certi ponuntur et quomodo de quarto, adhuc in-

cognito, dubitatur.

16. omnino *pro* ´etenim D.   numeri *om.* quoque *om.* D.

17. per alterum (alium D).   numerum *om.* D.

18. per.   erunt (+ 7 siue D).

19. numeri (*om.* D) almagul (maghulis D) id est incognitum.   Et huius (+ -modi D).

23. Almuszarar C.   videlicet *om.* D.   quo *om.*

24. inquiritur V.

25. Primum D.   duorum *om.* D.

26. in alterum (alium D).   fient.   Hunc ergo numerum per alterum (alium D) cognitum; Hunc C *sed del.*

27. Almuzahar + id est 32 per 10 diuide. Almuszarar C.   erunt V; exeund D.

28. qui *om.*   numeri.   designantes, qui videlicet eo numero per quem.   est *om.* D.

is indicated by the question as to the quantity. The rule of this kind of problem is to consider the three quantities which are given or known, of which two are always found to be inversely proportional to each other. These two are to be multiplied one by the other and the product of the multiplication is to be divided by the third number, likewise known, which is inversely proportional to the unknown. Now the quotient of this division will be the number which is sought, and it is inversely proportional to the number by which you divide. But lest some error be made in this type of problem we give an example of it.[1]

A Problem of the First Type

Ten for six, how many for four?

Observe now in what manner the given numbers are related, according to what we have said. For when you say "ten," you give the measure, and when you say "for six," you state the price. When you ask, "how much?" you give the unknown, called Almuhen or Magul, and saying "for four," you mention the cost. You note further that three of these, that is, 10, 6, and 4, are known and definite numbers, and the question is concerning the fourth or unknown number. If now you take account of the rule given, you multiply the first by the last, that is, 10 by 4, for they are the known and definite numbers which are inversely proportional to each other. It is necessary to divide the product, that is, 40, by the other known and definite number, that is, the measure, which is 6. This gives $6\frac{2}{3}$, designating the unknown number. This number is inversely proportional to the number six, which in Arabic is called Alszarar.

A Problem of the Second Type

An example of the second type of such problems is given by the question, "ten for eight, what is the cost of four?"

Ten now is the measure which is inversely proportional to the unknown cost, and eight designates the price which is inversely proportional to the quantity, 4. Therefore multiply one of the two known and inversely proportional numbers by the other, that is, 4 by 8, and you will have 32. Divide 32 by the third known number, 10, which is the measure. This gives $3\frac{1}{5}$, designating the cost which is inversely proportional to the number by which

---

[1] As we have noted in the introduction, page 44, this paragraph is not carefully translated by Robert of Chester, and he added the last sentence. The corresponding passage in the Libri version (*op. cit.*, pp. 268–269) is entitled, *Capitulum conventionum negociatorum*, and begins as follows: Scias quod conventiones negotiationis hominum, que sunt de emptione et venditione et cambitione et conductione et ceteris rebus, sunt secundum duos modos, cum quattuor numeris quibus interrogator loquitur.

Leonard of Pisa (*Liber abbaci*, p. 2 and p. 83) follows somewhat the terminology of this version in his discussion. The title (p. 2) is given: *De emptione et venditione rerum uenalium et similium.* The section opens (p. 83), as follows: 'In omnibus itaque negotiationibus quattuor numeri proportionales semper reperiuntur, ex quibus tres sunt noti, reliquus uero est ignotus.'

1   His igitur duobus modis omnia quae venalia dicuntur, absque omni errore possunt tractari, si deus voluerit.

Quaestio seu interrogatio vltima

Homo in vinea 30 diebus pro decem denariis conducitur, ex quibus operatus est
5 sex diebus, quantum ergo precii totius debet accipere?

Expositio huius est, quoniam manifestum est quòd sex dies quintam partem
totius temporis adimplent, et quòd hoc quod ex precio ei contingerit, sit secundum
quod ipse ex toto tempore, scilicet ex 30 diebus, sit operatus. Quod autem diximus, sic exponitur. Quoniam quando mensem id est 30 dies dixisti Almuzahar
10 protulisti; et quando 10 dixisti, Alszarar; quando verò 6 dies, Almuhen pronunciasti, quando deinde dixisti quantum precii ei contingerit, Alchemon nunciasti.
Multiplica ergo Alszarar, qui sunt 10, cum Almuhen qui ei opponitur, id est cum 6,
et quod ex multiplicatione excreuerit, 60 scilicet, in 30 Almuzahar diuide, et exeunt
2 denarii, et ipsi erunt Alchemon, id est pars quae homini contingerit.

15   Hoc igitur modo quicquid huius tibi propositum fuerit, ex rebus venalibus siue
ponderibus, seu ex omnibus quae ad haec attinent, agendum erit.

Laus deo praeter quem non est alius.

Finis libri restaurationis et oppositionis numeri quem Robertus Cestrensis de
Arabico in latinum in ciuitate Secobiensi transtulit, [Era] anno millesimo centesimo
20 octogesimo tertio.

1. omni *om.* D.   4. dragmata.
5. precium.
6. quod iam *pro* quoniam.   quoniam *pro*
quòd.   partem *om.*
7. mensis *pro* totius temporis.   quòd [1] *om.* D.
eius *pro* ei V.
8. ex mense.   scilicet ex 30 diebus *om.*
9. Almuszarat C.
10. dixisti 10 + dixisti.   et quando 6 dixisti
dies.
11. Et quando dixisti.
12. id est 10 V; 10 D.   6 + et fient 60.

13. 60 scilicet *om.*   super 30, id est super
Almusarar (Almuzaar D) et erunt duo dragmata et
ipsa.
14. erunt *om.* D.   contingit D.
15. Hoc ergo modo quotquot tibi huius positum
V; Huius ergo modi quicquid nisi positum D.
venalibus + ac aliis.
16. his V; hiis D *pro* ad haec.   est.
17–20. *om.* D.
17. Deo gracias V.
18. Explicit V.
19. Sectobiensi V.   Era M C lxxxiii V.

you divided. According to these two methods it is possible to treat all commercial problems, without error if God will.

The Last Problem or Question

A man is hired to work in a vineyard 30 days for 10 pence. He works six days. How much of the agreed price should he receive?

Explanation. It is evident that six days are one-fifth of the whole time; and it is also evident that the man should receive pay having the same relation to the agreed price that the time he works bears to the whole time, 30 days. What we have proposed, is explained as follows. The month, *i.e.* 30 days, represents the measure, and ten represents the price. Six days represents the quantity, and in asking what part of the agreed price is due to the worker you ask the cost. Therefore multiply the price 10 by the quantity 6, which is inversely proportional to it. Divide the product 60 by the measure 30, giving 2 pence. This will be the cost, and will represent the amount due to the worker.

This, then, is the method by which all proposed problems concerning commercial transactions or weights and measures and all related problems are to be solved.

Praise be to God, beside whom there is no other.

Here ends the book of restoration and opposition of number which in the year 1183 (Spanish Era) Robert of Chester in the city of Segovia translated into Latin from the Arabic.

1   REGULE 6 CAPITULIS ALGABRE CORRESPONDENTES

Prima. Quando numeris assimilantur ℞ committatur φ per ℞ et productum ostendit quesitum.

2^a. Quando φ assimilantur ʒ committatur φ per ʒ et radix propositi (producti) 5 ostendit quesitum.

3^a. Quando ℞ assimilantur ʒ committatur [per φ] per ʒ et productum ostendit quesitum.

4^a. Quando φ assimilantur ℞ ⁊ ʒ, φ ⁊ ℞ debent per ʒ committi; radix mediari, medium in se duci, productum numero addi. Radix tocius aggregati 10 minus medietate ℞ ostendit quod queritur.

5^a regula. Quando ℞ assimilantur φ ⁊ ʒ, ℞ ⁊ φ debent per ʒ committi; ℞ mediari, medium in se duci, a producto φ subtrahi, ℞ residui a medietate ℞ tolli et hiis residuum ostendit quesitum. Quod si ℞ residui a medietati ℞ subtrahi non potest, addere licet eandem.

15   6^a. Quando ʒ assimilantur φ ⁊ ℞. Hec debent per ʒ committi; ℞ mediari, medium in se duci, productum φ addi. Radix aggregati plus medietate ℞ ostendit quod queritur.

1–17. Regule . . . queritur *add* V.

RULES CORRESPONDING TO THE SIX CHAPTERS OF ALGEBRA

First. When roots are equal to a number, divide the number by (the number of) the roots, and the quotient represents the desired quantity.[1]

Second. When squares are equal to a number, divide the number by (the number of) the squares, and the root of that which you obtain represents the desired quantity.[2]

Third. When roots are equal to squares, divide (the number of) the roots by (the number of) the squares, and the quotient represents the desired quantity.[3]

Fourth. When a number is equal to the sum of squares and roots, divide by (the number of) the squares. Take one-half of (the number of) the roots after the division and multiply it by itself. To this product add the number. The root of this sum less one-half of the number of roots, represents that which is sought.[4]

Fifth. When roots are equal to a number and squares, divide the roots and the number by (the number of) the squares. Take one-half of (the number of) the roots after the division, and multiply it by itself. From this product subtract the number; the root of the remainder subtracted from one-half of (the number of) the roots is the desired quantity.[5] But if it is not possible [6] to subtract the root of the remainder from one-half of (the number of) the roots, it is permissible to add the same.

Sixth. When squares are equal to a number and roots, on this side divide by (the number of) the squares. Take one-half of (the number of) the roots after the division, and multiply it by itself. To this product add the number; the root of this sum plus one-half of (the number of) the roots represents that which is sought.[7]

---

[1] $ax = n$; $x = a/n$.

[2] $ax^2 = n$; $x^2 = a/n$; $x = \sqrt{a/n}$.

[3] $ax^2 = bx$; $x = b/a$.

[4] $ax^2 + bx = n$; $x = \sqrt{(b/2\,a)^2 + n/a} - b/2\,a$.

[5] $ax^2 + n = bx$; $x = b/2\,a \pm \sqrt{(b/2\,a)^2 - n/a}$.

[6] It is always possible to subtract the root of the remainder here from one-half of the number of the roots, even from the standpoint of the mathematician of the fourteenth century who made this table of rules, for the remainder would always be positive, since the square root of $(b/2\,a)^2 - n/a$ is always less than $b/2\,a$.

[7] $ax^2 = bx + n$; $x = b/2\,a + \sqrt{(b/2\,a)^2 + n/a}$.

The six rules with similar algebraic symbolism are found also in Codex Dresdensis, C. 80^m; see Wappler, *Programm, loc. cit.*, p. 11.

Combinationem triplicem in censibus reperimus, aut enim census et radices aequantur numeris, aut census et numeri aequantur radicibus, aut verò radices et numeri censibus aequantur.

5   *Exemplum combinationis primae*

Census et octo radices eius 33 denariis aequipollent. Quaeritur ergo, quantus est census?
                 Respondetur: 9.

Regula est, vt productum ex medietate radicum cum seipsa multiplicata, 10 scilicet 16, ad numerum, 33, addamus et resultant 49, cuius radix est 7. Atque ab hac radice medietatem radicum, 4 scilicet, subtrahamus, et manebunt 3, quae sunt radix census. Est igitur census 9.

Item si dicatur, duo census et octo radices sunt aequales quadraginta duobus denariis. Reduc quaestionem ad vnum censum sic. Si 2 census et 8 radices sunt 15 42 denarii, igitur vnus census et 4 radices sunt 21. Operando igitur per illud ex duplicatione et peruenies in fine ad intentum.

Item si dicas, medietas census et 4 radices sunt aequales $16\frac{1}{2}$ denariis.

Reduc quaestionem ad integrum censum, et dimidia totum, et patebit in fine intentum.

20   *Exemplum combinationis secundae*

Census et 15 denarii valent octo radices, quantus igitur est census?
                 Respondetur: 9, vel 25.

Regula. Duc medietatem radicum in se, et fiunt 16. Ex quibus subtrahe 15 denarios et manet vnum, cuius radicem, scilicet vnum, subtrahe a medietate radi- 25 cum, et manent 3, quae sunt radix census. Vel illud vnum adde ad medietatem radicum, et veniunt 5 quae sunt radix census. Patet igitur, quaestionem secundum vtranque partem esse veram, ideo oportet vt eas determines distinctiue.

                     Corollarium

Notandum si quadratum medietatis radicum minus fuerit quam est vltimus 30 numerus propositus, vt si dicatur, census et 15 numeri sunt aequales 6 radicibus, tunc quaestio impossibilis erit. Si verò fuerit vltimo proposito numero aequalis vt si dicatur, census et 9 denarii valent 6 radices, tunc medietas radicum est radix census. Si autem quaestio venerit cum pluribus censibus, aut paucioribus vno censu, reduc ad vnum censum, et operare vt supra dictum est.

35   *Exemplum combinationis tertiae*

Quatuor radices et 12 denarii censui aequipollent.

Reduc, hoc est multiplica, medietatem radicum in se, et fiunt 4. Quibus adde

SOME ADDITIONS IN EXPLANATION OF THE ALGEBRA

We find a triple combination of squares,[1] namely squares and roots equal to numbers, squares and numbers equal to roots, and finally roots and numbers equal to squares.

An Illustration of the First Type

A square and 8 of its roots equal 33 units. The question is, what is the square?[2]   Answer: 9.

The rule is to add to the number the product of half of the root multiplied by itself, that is 16 to 33, and 49 is obtained, of which the root is 7. Now from this root we subtract one-half of the number of the roots, namely 4, which leaves three. This is the root of the square, and the square is 9.[3]

Likewise if you are given that two squares and eight roots are equal to 42 units. Reduce in the following manner to one square. If two squares and eight roots are equal to 42 units, then one square and four roots are equal to 21. You operate therefore by duplication and arrive in the end at the desired result.[4]

Likewise if you are given one-half a square and four roots equal to $16\frac{1}{2}$ units. Reduce the problem to a whole square, take one-half of the whole, and in the end you arrive at the desired result.[5]

An Illustration of the Second Type

A square and fifteen units equal eight roots. What then is the square?
Answer: 9 or 25.

Rule. Multiply one-half of the roots by itself, and 16 is obtained. From this subtract the 15 units and one is obtained, of which you subtract the root, namely one, from one-half of the roots. This gives 3 as the root of the square. Or add that one to one-half of the roots and 5 is obtained as the root of the square. It is clear then that the problem is solved by each value, and hence it is necessary to determine separately both solutions.[6]

Corollary

It should be noted that if the square of half of the roots is less than the proposed number, as, for example, a square and the number 15 equal to 6 roots, then the problem is not solvable.[7] If the square of one-half of the roots is equal to the given number, as, for example, a square and 9 units equal to 6 roots, then the half of the roots is the root of the square. Moreover, if a problem is proposed involving more than one square or less than a whole square, reduce to one square, and operate as indicated above.

---

[1] Scheybl changes from *substantia* to *census;* see p. 69, n. 1.

[2] *Textus* in the margin precedes the statement of each problem, and *Minor* the solution.

112 denarios et veniunt 16, cuius radicem 4 adde ad medietatem radicum, et resultant 6, quae sunt radix census. Omne etiam, quod aut maius est vno censu aut minus, reducas ad vnum censum, atque deinde operare vt dictum est.

Proponit causas trium combinationum

5   Causa combinationis primae est haec:

Sit quadratum *a b* quod censum significet, sitque huius quadrati latus ipsius census radix. Et quoniam ductus vnius lateris in numerum radicum, quantitas

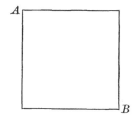

radicum sumptorum existit. Applicetur igitur ad vnumquodque latus quadrati quarta pars numeri radicum, scilicet duae radices lateri vni, et duae alii, et ita
10 deinceps. Resultat autem sic quadratum aliud, praeter quatuor eius angularia quadrata, quorum singulorum vnumquodque latus duo, hoc est quartam partem numeri radicum, continebit. Ex ductu igitur quartae partis numeri radicum in se quater, resultabunt illa quatuor parua angularia quadrata, atque haec eadem

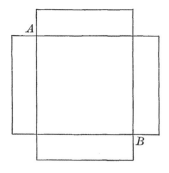
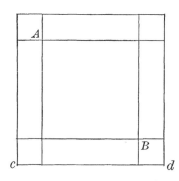

resultant ex multiplicatione medietatis numeri radicum in se semel. Si igitur
15 productum ex medietate numeri radicum in se multiplicata, addatur ad censum et ad radices, quadratum cuius vnius lateris quantitas aequalis est lineae *c d* describetur. Hoc autem latus excedit latus census in medietate numeri radicum, quia per quartam partem in vno extremo, et per quartam partem in altero huius census latus auctum est. Ideo subtracta medietate numeri radicum a tota linea
20 *c d* manebit quantitas radicis census quae quaerebatur.

Illa probatio procedit ex propositione quarta secundi Euclidis.

An Illustration of the Third Type

Four roots and twelve units are equal to a square.

Find, that is to say multiply, half of the roots by itself, and four is obtained. To this add the 12 units, and 16 is obtained. Add the root of this, 4, to one-half of the roots and 6 appears as the root of the square. Also whatever is given either greater or less than one square, reduce to one square and operate as indicated above.[1]

p. 130

The Explanation of the Three Types

The explanation of the first type is as follows: let the square AB represent x^2. Then the side of this square will represent the root of x^2, or x. When one side of this square is multiplied by the number of the roots, the quantity of the assumed roots is represented. Hence let there be applied to each side of the square the fourth part of the number of the roots, namely two roots to one side, two to another, and so on. Thus another square appears, lacking only four small corner squares of which each has every side equal to two, that is the fourth part of the number of the roots. Therefore by multiplying the fourth part of the number of the roots by itself four times, these four small corner squares are obtained, and this same result is obtained by the multiplication of half of the roots by itself once. If then the product of half the roots multiplied by itself be added to the square and roots, the square is formed of which one side is equal in value to the line cd. Moreover, this side exceeds the side of the unknown square by the half of the roots, since the side of the unknown square has been extended by the fourth part at one extremity and by the fourth part at the other. Hence by subtracting the half of the number of the roots from the whole line cd the value appears of the root of the square which is sought.

The proof of this follows from the fourth proposition of the second book of Euclid.[2]

[3] $x^2 + 8x = 33$; $x^2 + 8x + 16 = 49$; $x + 4 = 7$; $x = 3$; $x^2 = 9$.

[4] $2x^2 + 8x = 42$; $x^2 + 4x = 21$; $x + 4x + 4 = 25$; $x + 2 = 5$; $x = 3$.

[5] $\frac{1}{2}x^2 + 4x = 16\frac{1}{2}$; whence $x^2 + 8x = 33$, as above.

[6] $x^2 + 15 = 8x$; $x^2 - 8x + 16 = 1$; $x - 4 = \pm 1$; $x = 3$ or 5. Both roots are positive, and so two solutions are given.

[7] $x^2 + n = bx$; $x = b/2 \pm \sqrt{(b/2)^2 - n}$.

The roots are imaginary if $(b/2)^2$ is less than n, as in the illustration which Scheybl gives; similarly, the two roots are equal to each other and each equal to one-half the coefficient of the roots if the square of this quantity is the same as the given constant, that is $(b/2)^2 = n$.

p. 130

[1] $4x + 12 = x^2$; $x^2 - 4x + 4 = 16$; $x - 2 = 4$; $x = 6$.

[2] Euclid II, 4, following Heath, *The Thirteen Books of Euclid's Elements:* "If a straight line be cut at random, the square on the whole is equal to the squares on the segments and twice the rectangle contained by the segments." In modern notation, $(a + b)^2 = a^2 + b^2 + 2ab$.

Aliter sic

Sit census quadratum *a b*, cuius vni laterum applicetur medietas numeri radicum,
et alteri eius lateri applicetur medietas numeri radicum altera, sit autem primum
additum *a c*, secundum vero *a d*. Ad perficiendum igitur quadratum *c d*, deficit
5 quadratum quod vocetur *b e*, cuius latus medietati numeri radicum est aequale.
Patet igitur causa operationis; nam ex ductu medietatis numeri radicum in se,
resultat quadratum *b e*, et residuum est census cum adiectis radicibus.   Vnde cum
illud magnum quadratum notum sit, et eius radix nota, per subtractionem radicis
quadrati *b e* à radice quadrati *a e*, necessario radix census manebit.
10   Sequuntur figurae geometricae.

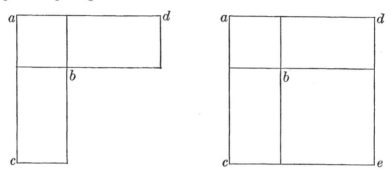

Exemplum combinationis secundae

Sit census quadratum *a b c d*, cuius lateri *b c* applicabo parallelogrammum *c b e f*,
et ponam ipsum 15. Totum igitur parallelogrammum *d a e f* est census et 15
denarii, et continet octo radices census.   Diuidam ergo lineam *d f* aequaliter in
15 punctum *g* et erigam super latus *g d* quadratum *g k m d*, et protraham *c b* vsque ad *l*
et ponatur litera *h* in locum vbi *g k* secat *a e* lineam.   Et quoniam *d b* est quadratum, cum ideo ex structura et propositione sexta secundi libri Euclidis, *b k* quadratum sit, duo parallelogramma *g b* et *b m* inter se aequalia erunt; mox deinde, per
communem quandam noticiam, *g a* et *c m* aequalia.   Atque tandem cum *g a* ex
20 propositione 36 libri primi Eucli[dis], *g e* parallelogrammo aequale sit, *c m* eodem
parallelogrammo *g e* aequale erit.   Subtractis igitur quindecim, hoc est gnomone
l d h à quadrato *g m* quod est sedecim, manet vnitas quadratum *b k*. Et quia
notum, latus igitur vel radix, quae est linea *l b*, erit nota.   Sed et tota *l c* nota est,
manet igitur, post subtractionem *l b* ab *l c* linea, et *b c* linea nota.   Et in hoc casu,
25 quo punctus *g* cadit in parallelogrammum applicatum censui, sequuntur figurae
geometricae.

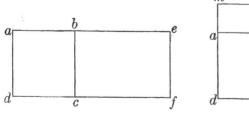
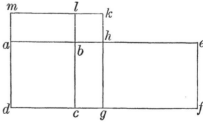

Another Demonstration

Let the square ab represent x^2. To the side of it apply one-half the given number of roots and to another side of it apply the other half of the number of the roots. Let the first addition be represented by ac and the second by ad. To complete the square cd the square called be is lacking and the side of this is equal to one-half the number of the roots. The reason for our procedure (in halving the roots) is now apparent; for by multiplying the half of the number of roots by itself we obtain the square be, and the remainder is x^2 together with the added roots. Now since the larger square is known, and the root of it is known, then by subtracting the root of the square be from the root of the square ae, necessarily the root of our unknown square remains.

The geometrical figures follow.[1]

An Illustration of the Second Type

Let the square abcd represent x^2, and to its side bc apply the parallelogram cbef, which we take to be equivalent to 15 units. Then the whole rectangle daef is equivalent to x^2 and 15 units, thus containing 8 roots of x^2 (8 x). Divide now the line df into two equal parts by the point g and erect a square gkmd upon the side gd. Extend cb up to a point l (the point of intersection with the upper side of the square upon gd). Mark the point of intersection of gk with ae by the letter h. Since db is a square, it follows by our construction and by proposition 6 of the second book of Euclid [2] that bk is a square and further that the two parallelograms gb and bm are equal to each other.[3] Further, then, by a well-known axiom [4] ga and cm are equal. Also finally, since ga is equal to the parallelogram ge by the thirty-sixth proposition of the first book of Euclid,[4] it follows that cm is equal to the same parallelogram ge. Hence subtracting 15, that is the gnomon ldh, from the square gm, which is 16, one is left as the area of the square bk. Since the area is known, the side or root, which is the line lb, is also known. But the whole line lc is known, whence it follows by subtraction of lb from lc that the line bc is known. In this instance it is to be noted that the point g falls within the parallelogram which is applied to the square.

The geometrical figures follow.

[1] Compare these and subsequent figures with the corresponding figures on pages 77–89.

[2] Euclid II, 6, following Heath, *The Thirteen Books of Euclid's Elements*, p. 1. "If a straight line be bisected and a straight line be added to it in a straight line, the rectangle contained by the whole with the added straight line and the added straight line together with the square on the half is equal to the square on the straight line made up of the half and the added straight line." In modern algebraic notation, $(a + x) x + (a/2)^2 = (a/2 + x)^2$.

[3] By Euclid I, 43, following Heath, *loc. cit.*, "In any parallelogram the complements of the parallelograms about the diameter are equal to one another."

[4] Euclid I, 36, following Heath, *loc. cit.*, "Parallelograms which are on equal bases and in the same parallels are equal to one another."

1   Sed esto iam, quod punctus *g* cadat in latus census vel quadrati.

Sit census vt prius *a b c d*, et parallelogrammum, numero quindecim, lateri *b c* applicatum *c b e f*. Tunc super medietatem lineae *d f*, erigo quadratum *g k m f* et accipio lineam *c o* aequalem lineae *c f* et protraho lineam *n h o p* aequedistantem 5 lineae *d f*. Et quoniam lineae *d c* et *c b* inter se aequales sunt, lineae quoque *c o* et *c f* aequales; et parallelogrammum igitur *d o* propter aequalitatem linearum, parallelogrammo *c e* aequale erit. Sed quia aequalia etiam inter se sunt ex propositione 43 primi libri Euclidis, duo supplementa *g o* et *o m*, cum per subtractionem aequalium ab aequalibus, ex communi quadam noticia, aequalia relinquantur, 10 parallelogrammum *d h* parallelogrammo *l e* cum quadrato *c p* aequale erit. Sed quia *d h* aequale etiam est, ex propositione 36 primi, parallelogrammo *g o* cum quadrato *c p*; haec igitur duo, *g o* parallelogrammum et *c p* quadratum, prioribus duobus *l e* parallelogrammo et *c p* quadrato ex communi quadam noticia, aequalia. Atque mox, per ablationem quadrati *c p* communis, *g o* ipsi *l e* parallelogrammo 15 aequale. Sunt vero *c o* et *l m* lineae inter se aequales. Ergo et linea *m e* lineae *g c* aequalis erit, atque ita aequalis etiam lateri quadrati *h l*. Cum ergo quadratum *g m* sit notum, eo quòd eius latus sit medietas. numeri radicum, si subtrahantur ab eo 15, quae gnomonem valent, quadrato *h l* circumiacentem et cui etiam parallelogrammum *d o* aequale est, manebit quadratum *h l* notum. Vnde et latus eius 20 notum. Sed quia illud est aequale lineae *g c* vel *m e*, latere igitur quadrati *h l* ad medietatem numeri radicum *m f*, latus quadrati *g m* addito, constituitur linea *e f* nota. Atque haec est radix census. Patet ergo propositum.

Sequuntur figurae geometricae.

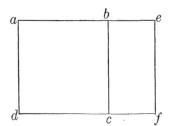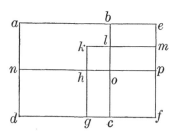

Exemplum combinationis tertiae

25   Quatuor radices et duodecim denarii censui aequipollent.

Causa huius combinationis est haec. Sit census quadratum *a b c d* ignotum continens quatuor radices et 12 denarios. Ex isto igitur quadrato resecabo parallelogrammum *b c e f* continens quatuor radices; manebit ergo parallelogrammum *a e* continens precise duodecim denarios. Deinde diuidam lineam *e c*, 30 numerum radicum, aequaliter in puncto *g*, et super latus *e g* erigam quadratum *g e h k*. Super latus etiam *d g* erigam quadratum *g d m l*, et secet linea *m l* lineam *e f* in puncto *n*. Et quoniam lineae *a d* et *d c* inter se aequales sunt, linea quoque *m d* et *d g* aequales, cum per subabiectionem aequalium ab aequalibus, linea *a m* lineae *g c* ex communi quadam noticia, aequalis sit. Eadem *a m* linea propter aequa- 35 litatem, ex altera quadam noticia, lineae *e g*, atque mox etiam lineae *h k*, aequalis erit. Item, quia lineae *h e* et *e g* inter se aequales sunt, lineae quoque *n e* et *d g*

Now let the point *g* fall within the side of the square.

Let the square as before be represented by *abcd* and the parallelogram *cbef*, equivalent to the given number 15, be applied to the side *bc*. Then upon half of the line *df* erect the square *gkmf*. Construct the line *co* equal to the line *cf* and draw the line *nhop* everywhere equally distant from the line *df*. Since the lines *dc* and *cb* are equal to each other (being sides of a square), and further, the lines *co* and *cf* equal (by construction), it follows that the parallelogram *do* is equal to the parallelogram *ce* on account of the equality of the sides. But further, since the two supplementary rectangles *go* and *om* are equal to each other by the forty-third proposition of the first book of Euclid, by subtracting equals from equals according to the well-known axiom, the remainders are equal, giving the parallelogram *dh* equal to the parallelogram *le* together with the square *cp*. But also since *dh* is equal to the parallelogram *go* plus the square *cp* by the thirty-sixth proposition of the first book (of Euclid), it follows that these two, the parallelogram *go* and the square *cp*, are equal to the preceding two, the parallelogram *le* and the square *cp* by another well-known axiom.[1] Whence by subtracting the common part, the square *cp*, *go* is equal to the parallelogram *le*; and indeed the lines *co* and *lm* are also equal to each other. Further, the line *me* will be equal to *gc*, and so also equal to the side of the square *hl*. Since the square *gm* has a known area, by the fact that its side is one-half of the given number of roots, if from it there be subtracted 15 which is represented by the gnomon *gfm*, circumscribed about the square *hl* and equal, as we have just shown, to the parallelogram *do*, the square *hl* remains known in area; whence also the side of it is known. But since this side of the square *hl* is equal to the line *gc* or *me*, when added to the half of the number of the roots *mf*, a side of the square *gm*, the line *ef* is then known. And this is the root of the unknown square. The proposed question is solved.

The geometrical figures follow.

An Illustration of the Third Type

Four roots and twelve units are equal to x^2.

The explanation of this type is as follows. Let x^2 be represented by the unknown square *abcd* which contains 4 roots and 12 units. From this square cut off the parallelogram *bcef* containing four roots. The parallelogram *ae* consequently will contain precisely 12 units. Then divide the line *ec*, representing the number of the roots, into two equal parts by the point *g* and upon the side *eg* erect the square *gehk*. Also upon the side *dg* erect a square *gdml*, and let the line *ml* cut the line *ef* in the point *n*. Since the lines *ad* and *dc* are equal to each other, and further the lines *md*

[1] Following Heath, *loc. cit.*, "Things which are equal to the same thing are also equal to one another."

1 aequales, cum per subtractionem aequalium ab aequalibus, linea *n h* lineae *d e* aequalis sit. Eadem *n h* linea, propter aequalitatem, lineae *m n* aequalis erit. Igitur superficies *h l* et *m f* aequales. Duae igitur superficies *d n* et *h l* simul sumptae, ex communi illa noticia, si aequalibus aequalia adiiciantur, vni superficiei 5 *d f* quae 12 continet, aequales erunt. Vnde si numerus 12 addatur ad quadratum *e k*, cuius latus vel radix est linea *e g*, medietas numeri radicum, resultabit quadratum *d l*, cuius latus vel radix est linea *d g*. Si igitur illi radici addatur medietas numeri radicum, quae est linea *g c*, linea *d c*, quae est latus census, proueniet. Patet ergo propositum.

10   Sequuntur figurae geometricae.

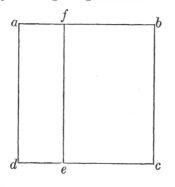 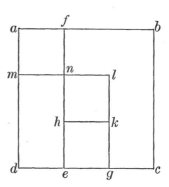

Vel sub alia forma, sic :

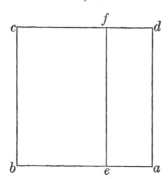 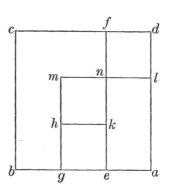

Sequuntur multiplicatio cum additis et diminutis

Sicut idem est multiplicare compositum cum composito, et multiplicare vtramque partem compositi cum vtraque parte compositi, vt sic fiat quadruplex multi-15 plicatio, scilicet articuli cum articulo et digiti cum articulo, deinde articuli cum digito, et quarto digiti cum digito. Comsimiliter potest fieri quadruplex multiplicatio, vbi articulus et digitus cum articulo praeter digitum multiplicare debet, et similiter vbi articulus praeter digitum cum articulo et digito, vel contrà, multiplicari debet. Vnde indicimus quadruplicem multiplicationem, atque in omni multi-20 plicatione tali praedicta haec regula notanda est.

Si digiti tam in multiplicando quam in multiplicante additi cum articulis aut

and *dg* are equal, then by subtracting equals from equals the line *am* will be equal to the line *gc* by this well-known axiom. On account of this equality the same line *am* will, by another axiom,[1] be equal to the line *eg* and hence also to *hk*. Further, since the lines *he* and *eg* are equal to each other and also the lines *ne* and *dg* are equal, then by subtracting equals from equals the line *nh* is equal to the line *de*. On account of this equality the same line *nh* will be equal to the line *mn*. Hence the areas *hl* and *mf* are equal. Therefore the two areas *dn* and *hl* taken together, by the well-known axiom "if equals be added to equals," will be equal to the single area *df*, which contains 12 units. Whence if the number 12 be added to the square *ek*, whose side or root is the line *eg*, the half of the number of the roots, the square *dl* is obtained whose side or root is the line *dg*. If, then, the half of the number of the roots, which is represented by the line *ge*, be added to that side (*dg*) the line *dc* will be obtained, which is the side of the unknown square. The proposed problem is solved.

The geometrical figures follow.

[Or in another form, thus]

Multiplication with Positives and Negatives [2]

Since the multiplication of a composite [3] number by a composite number is the same as the multiplication of each part of the one composite by each part of the other, so it follows that the multiplication is fourfold, namely article by article, digit by article, then article by digit, and fourthly digit by digit. Similarly, you may have a fourfold multiplication when an article and a digit is to be multiplied by an article less [4] a digit, or the reverse.[5] It seems desirable to indicate the fourfold nature of this multiplication (by some examples), and in every multiplication of this kind this rule is to be followed.

If the digits in the multiplicand as well as in the multiplier are added to

---

[1] The halves of equals are equal, and the axiom of the preceding note.

[2] Attention is called to a similar use of this expression in the text (p. 32). Johannes de Muris in the third book of the *Quadripartitum numerorum*, doubtless familiar to Scheybl, entitles a similar chapter, *De multiplicatione et diuisione additorum et diminutorum* (Cod. Pal. Vind. 4770, fol. 230 [6]); see also Karpinski, *The "Quadripartitum numerorum" of John of Meurs, loc. cit.,* p. 110.

[3] In this discussion Scheybl uses the more common terms such as "composite numbers," "articles," and "digits," instead of "nodes" and "units" as in the corresponding section of Robert of Chester's text (pp. 88–96).

[4] *praeter* is used to suggest the idea of negative, illustrating in fact the Arabic conception of negative, namely that the "article" or ten (in this instance) is incomplete by the digit which is subtracted from it.

[5] Probably including all four types, $(x + a)(x + b)$, $(x + a)(x - b)$, $(x - a)(x + b)$, and $(x - a)(x - b)$. That multiplication is commutative was doubtless felt, even though not expressed.

1 diminuti fuerint in vtroque, tunc multiplicatio quaelibet debet addi. Si autem
vnus fuerit additus et alter diminutus, tunc ista multiplicatio subtrahi debet a
productis.

Sequitur huius rei vel regulae exemplum

5 Idem est multiplicare 8 cum 17, et multiplicare 10 praeter 2 cum 20 praeter 3,
et hoc fiet sic. Multiplica 10 cum 20, et resultabant 200. Deinde multiplica
praeter 2 cum 20, et resultabunt 40, subtrahenda à 200 et manent 160. Tertio
multiplica 10 cum praeter 3, et resultabunt 30, subtrahenda a 160 et manent 130.
Quartò multiplica praeter 2 cum praeter 3 et resultabunt 6 addenda, et veniunt
10 136. Igitur octo decies septies sunt 136.

Sequitur huius rei calculus.

$$\begin{array}{r} 10 \text{ praeter } 2 \\ \underline{20 \text{ praeter } 3} \\ 200 \text{ praeter } 40 \\ \underline{\text{praeter } 30 \text{ et } 6} \\ 136 \text{ et caet.} \end{array}$$

15

Aucta minuta simul minues; sed caetera iunges.
Aequantur numero radices censibus ambo,
In medio minues, alibi quod colliges addes.

20   *Alia tria carmina, quae trium aequationum exempla proponunt*

Census et 8 res 30 valent simul et tres.
Census cum seno, res quinque valere notato.
Aequiualent censum res 4 et duodenus.

Sequuntur nunc in declaratione exempla multiplicationis alia

25 Si cum 10 praeter rem debeas multiplicare 10.
Multiplica 10 cum 10, et fiunt 100. Deinde multiplica 10 cum re, et fiunt 10
res diminuendae. Dic ergo quòd resultent 100 denar. praeter 10 res.
Item secundò: si cum 10 et re multiplicares deberes 10.
Multiplica 10 cum 10 et re, et resultabunt 100 denarii et 10 res.
30 Item tertiò. Si cum 10 et re multiplicare deberes 10 et rem.
Multiplica 10 cum 10, et fiunt 100. Deinde multiplica rem cum 10, et fiunt 10
res addendae. Tertiò multiplica 10 cum re, et fiunt 10 res addendae. Quartò
multiplica rem cum re, et fit census addendus. Erit ergo totum 100 dena., 20
res et census.
35 Item quartò. Si cum 10 praeter rem multiplicare debes 10 praeter rem.
Multiplica 10 cum 10, et fiunt 100. Deinde multiplica rem diminutam cum 10,
et fiunt 10 res diminuendae. Manent autem 100 praeter 10 res. Tertiò multiplica 10 cum re diminuta, et fiunt 10 res diminuendae. Manent autem 100 praeter
20 res. Quartò multiplica rem diminutam cum re diminuta, et fit census addendus.
40 Totum igitur erunt 100 denarii, census praeter 20 res.

the articles, or both negative, then the fourth product is positive. If, however, one term in one binomial is positive and the other corresponding term is negative then this product should be subtracted from the other products.

There Follows an Illustration of this Rule

To multiply 8 by 17 is the same as to multiply 10 less 2 by 20 less 3, and it will be done in this way. Multiply 10 by 20, giving 200; then multiply negative 2 by 20, giving 40, which subtracted from 200 leaves 160; in the third place multiply 10 by negative 3, giving 30, which subtracted from 160 leaves 130; in the fourth place multiply negative 2 by negative 3, giving positive 6, which being added gives 136. Therefore 8 times 17 gives 136.

The calculation follows:

$$
\begin{array}{r}
10 - 2 \\
\underline{20 - 3} \\
200 - 40 \\
\underline{- 30 + 6} \\
136, \text{ etc.}
\end{array}
$$

Positives by negatives you subtract, but other products you add. When roots are equal to both number and squares, the half (is multiplied by the half) and you subtract, otherwise you add that which you obtain.[1]

Three other lines of verse which illustrate the three types of equations.

A square and 8 roots equal 33.

A square together with 6 equals 5 roots.

Four roots and 12 are equal to a square.

In further explanation: other examples of multiplication.

Suppose you are to multiply 10 by 10 less x.

Multiply 10 by 10, giving 100. Then multiply 10 by (negative) x, giving negative 10 x. Hence the product is 100 units less 10 x.

Second. Suppose you are to multiply 10 by 10 plus x.

Multiply 10 by 10 and by x, giving 100 units and 10 x.

Third. Suppose you are to multiply 10 plus x by 10 plus x.

Multiply 10 by 10, giving 100. Then multiply x by 10, giving positive 10 x. Thirdly, multiply 10 by x, giving positive 10 x. Fourthly, multiply x by x, giving positive x^2. Hence the product is 100 units, 20 x and x^2.

Fourth. Suppose you are to multiply 10 less x by 10 less x.

Multiply 10 by 10, giving 100. Then multiply negative x by 10, giving negative 10 x. The total, so far, is 100 less 10 x. Thirdly, multiply 10 by negative x, giving negative 10 x, or, in all, 100 less 20 x. Fourthly, multiply negative x by negative x, giving x^2 positive. The final product will be 100 units, and x^2 less 20 x.

[1] These verses are somewhat obscure. The meaning probably is that in the type $x^2 + n = bx$ the number is subtracted from the square of half the coefficient of x, whereas in the other two types of complete quadratics you subtract the number from this square.

1   Item quintò. Si cum 10 praeter rem debes multiplicare 10 cum re.

Multiplica 10 cum 10, et fiunt 100. Deinde multiplica rem cum 10, et fiunt 10 res addendae. Tertiò multiplica 10 cum praeter rem, et fiunt 10 res diminuendae. Quartò multiplica rem adiectam cum re diminuta, et sit census diminu-
5 tus. Est ergo totum 100 dena. praeter censum.

Item sextò. Si cum 10 praeter rem deberes multiplicare rem.

Multiplica rem cum 10, et fiunt 10 res; deinde multiplica etiam rem cum re diminuta, et fit census diminutus. Est igitur totum, 10 res praeter censum.

Item septimò. Si cum 10 et re deberes multiplicare rem praeter 10.

10 Multiplica rem cum 10, et fiunt 10 res. Deinde multiplica praeter 10 cum 10, et fiunt 100 diminuenda. Tertiò multiplica rem cum re, et fit census addendus. Quartò multiplica praeter 10 cum re, et fiunt 10 res diminuendae. Est igitur totum, census praeter 100 denar.

Item octauò. Si cum 10 et medietate rei multiplicare deberes medietatem
15 denarii praeter quinque res.

Multiplica medietatem denarii cum 10, et fiunt 5. Deinde multiplica praeter 5 res cum 10, et fiunt 50 res diminuendas. Tertiò multiplica medietatem denarii cum medietate rei, et fit quarta rei addenda. Quartò multiplica praeter 5 res cum medietate rei, quod est multiplicare duas res et dimidiam cum vna re, et fiunt duo
20 census et $\frac{1}{2}$ diminuendi. Est igitur totum, 5 denarii praeter 2 census et $\frac{1}{2}$ et praeter 49 res et $\frac{3}{4}$ rei.

Item nono. Si cum vno denario praeter sextam denarii, multiplicare deberes denarium praeter sextam.

Multiplica denarium cum denario, deinde praeter sextam denarii cum denario,
25 et fit denarius praeter sextam denarii. Tertiò multiplica denarium cum praeter sextam denarii, et fit vna sexta minuenda; manent ergo quatuor sextae. Quartò duc vel multiplica praeter sextam cum praeter sextam denarii, et fit vna trigesima sexta addenda. Est igitur totum viginti quinque trigesimae sextae vnius denarii.

Item decimò. Si cum 10 et re multiplicare deberes 10 et rem praeter 10. Idem
30 est ac si cum 10 et re multiplicare deberes rem.

Et tunc fit totum, census et 10 res.

De radicum duplatione, triplatione et quadruplatione

Notandum quod cum census radicem, siue notam siue surdam duplare volueris, multiplica duo cum duobus, et cum producto multiplica censum. Et erit huius
35 producti radix, dupla radix census propositi quae quaerebatur. Consimiliter si eius triplum habere volueris, multiplica ternarium cum ternario. Et si eius quadruplum, multiplica quatuor cum quatuor, et ita deinceps, et cum producto multiplica censum propositum, et erit producti radix, census propositi radix duplata, triplata vel quadruplata et caet.

40   ### Idem in fractionibus obseruandum

· Vt si medietatem radicis habere consideras, multiplica medietatem cum medietate, et cum producto deinde censum, et tunc radix producti ostendit quaesitum. Similiter si tertiam partem habere volueris, multiplica tertiam partem cum tertia parte. Et ita deinceps.

Fifth. Given to multiply 10 plus x by 10 less x.

Multiply 10 by 10, giving 100. Then multiply x by 10, giving positive 10 x. Third, multiply 10 by negative x, giving negative 10 x. Fourth, multiply the positive x by the negative x, giving negative x^2. The sum total is 100 units less x^2.

Sixth. Given to multiply x by 10 less x.

Multiply x by 10, giving 10 x; then multiply x by negative x, giving negative x^2. The total is therefore 10 x less x^2.

Seventh. Given to multiply x less 10 by 10 plus x.

Multiply x by 10, giving 10 x. Then multiply negative 10 by 10, giving negative 100. Third, multiply x by x, giving positive x^2. Fourth, multiply negative 10 by x, giving negative 10 x. The total is therefore x^2 less 100 units.

Eighth. Given to multiply one-half a unit less 5 x by 10 plus $\frac{1}{2}x$.

Multiply the half unit by 10, giving 5. Then multiply negative 5 x by 10, giving negative 50 x. Third, multiply the half unit by $\frac{1}{2}x$, giving $\frac{1}{4}x$ positive. Fourth, multiply negative 5 x by $\frac{1}{2}x$, which is the same as multiplying $2\frac{1}{2}x$ by x, giving $2\frac{1}{2}x^2$ negative. The total therefore is 5 units less $2\frac{1}{2}x^2$, and less $49\frac{3}{4}x$.

Ninth. Given to multiply a unit less $\frac{1}{6}$ by a unit less $\frac{1}{6}$ of a unit.

Multiply a unit by a unit, then negative $\frac{1}{6}$ of a unit by a unit, giving a unit less $\frac{1}{6}$ of a unit Third, multiply a unit by negative $\frac{1}{6}$ of a unit, giving negative $\frac{1}{6}$ of a unit. These give $\frac{4}{6}$ of a unit. Fourth, multiply negative $\frac{1}{6}$ by negative $\frac{1}{6}$ of a unit, giving positive $\frac{1}{36}$. The total is therefore twenty-five thirty-sixths of a unit.

Tenth. Given to multiply 10 plus x less 10 by 10 plus x. This is the same as to multiply x by 10 plus x. Hence the product is x^2 plus 10 x.

Concerning the doubling, tripling, and quadrupling of radicals

It should be noted that when you wish to double the root of a square, either a definite root or a surd, you multiply 2 by 2 and multiply the given square by this product. The root of this product will be the double which you seek of the root of the proposed square. Similarly if you wish its triple, you multiply 3 by 3. And if you wish the quadruple of it, you multiply 4 by 4, and so on; and finally you multiply the proposed square by the product; the root of the product thus obtained will be the double, triple, or quadruple, etc., of the proposed square.

A similar note on fractions

If you have it in mind to obtain half of the root, multiply one-half by one-half, then the given square by the product. The root of this final product gives the desired result. Similarly if you wish to have a third part, multiply a third part by a third part, and so on.

1

De radicum diuisione

Nota, si radicem nouenarii in radicem numeri 4 diuidere volueris, diuide 9 in 4 et erit exiens 2 et quarta, atque exeuntis huius radix, quae est vnum et semis, erit numerus exiens diuisionis radicis in radicem. Quòd si duas radices nouenarii in
5 radicem numeri 4 diuidere volueris, quaeras primò duplum radicis nouenarii secundum quod docuimus, et illud diuide in radicem numeri quatuor.

De radicum multiplicatione

Si radicem nouenarii cum radice numeri 4 multiplicare volueris, multiplica 9 cum 4, et producti radix est quod quaeris. Ita cum aliis.
10 Quòd si radicem tertiae cum radice medietatis multiplicare volueris, multiplica tertiam cum medietate, et producitur sexta, cuius radix est quod quaeris.

Quod si duas radices nouenarii cum tribus radicibus quaternarii multiplicare volueris, inquiras primò secundum quod supra docuimus censum cuius radix est duplum radicis numeri 9. Consimiliter inquiras censum cuius radix est triplum
15 radicis numeri 4, et multiplica vnum censum cum altera et radix producti erit quaesitum.

Sequuntur nunc quatuor aenigmata

Primum

Radix ducentorum subtractis 10, addita ad duplum subtracti, scilicet ad 20,
20 subtracta ducentorum radice aequaliter erit 10.

Dicit aenigma, si a radice numeri 200 subtrahantur 10, id deinde quod relinquitur ad 20 addatur, ab hoc collecto postea radix numeri 200 auferatur, quòd subtractum tandem, hoc est 10, aequaliter maneant. Hoc autem sic probabitur. Sit linea *a b* radix ducentorum, a qua resecabo lineam *a c*, quae sit 10. Deinde lineae
25 *a b* adiungam *b d*, lineam quae sit 20, a qua resecabo lineam *b e*, aequalem lineae *a b*, et a linea *b e* resecabo lineam *b f* aequalem lineae *a c*. Erit igitur linea *c b* aequalis lineae *f e*. Est autem linea *c b* radix ducentorum exceptis 10. Ac linea *e d*, 20 excepta radice ducentorum, et *c b* linea est aequalis *f e* lineae, igitur linea *f d* radix ducentorum erit exceptis 10, addita ad 20 excepta radice ducentorum.
30 Quòd autem haec linea *f d* sit praecise 10, probabo. Linea *b d* est 20 et cum *b f* linea aequalis sit lineae *a c*, quae 10 posita est, oportet igitur quod et *f d* linea propter aequalitatem 10 sit, quod erat probandum.

Sequitur figura.

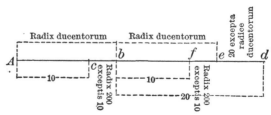

On the division of radicals

Notice, if you wish to divide the root of 9 by the root of 4, divide 9 by 4, giving $2\frac{1}{4}$. Of the result take the root, which is $1\frac{1}{2}$, and the resulting number will be the quotient of the root divided by the root. But if you wish to divide two roots of 9 by the root of the number 4, you seek first the double of the root of 9 according to that which we have explained, and divide the product by the number 4.

On the multiplication of radicals

If you wish to multiply the root of the number 9 by the root of the number 4, multiply 9 by 4 and the root of the product is that which you seek. Other multiplications are similar.

Thus if you wish to multiply the root of $\frac{1}{3}$ by the root of $\frac{1}{2}$, you multiply $\frac{1}{3}$ by $\frac{1}{2}$, giving $\frac{1}{6}$, and the root of this is that which you seek.

In the same way if you wish to multiply two roots of 9 by three roots of 4, you first try to find, as we have explained above, the square whose root is twice the root of the number 9. Similarly you try to find the square whose root is three times the root of the number 4, and you multiply the one square by the other. The root of the product will be that which you seek.

Now follow four problems [1]: First problem

The root of 200 less 10, added to the double of that which is subtracted, in other words to 20, will be equal to 10 when the root of 200 is subtracted.[2]

The problem says that if from the root of the number 200, 10 is subtracted, and if then 20 is added to that which remains, and if from this sum the root of the number 200 is taken away, we finally have left that which was subtracted, namely 10. This will be proved in the following manner:

Let $a\,b$ represent the root of 200, and from it cut off the line $a\,c$, representing 10. Then to the line $a\,b$ join $b\,d$, a line which is 20 (units in length). From this cut off $b\,e$, equal to the line $a\,b$, and from $b\,e$ cut off the line $b\,f$ equal to the line $a\,c$; the line $c\,b$ will be equal to the line $f\,e$. Moreover the line $c\,b$ is the root of 200 less 10, and the line $e\,d$ is 20 less the root of 200. The line $c\,b$ is equal to the line $f\,e$, whence the line $f\,d$ equals the root of 200 less 10 plus 20 less the root of 200. Moreover I shall prove that this line $f\,d$ is exactly 10. The line $b\,d$ is 20, and as $b\,f$ is equal to the line $a\,c$, which was taken as 10, it follows then that the line $f\,d$, on account of this equality, is 10, which was to be proved.

---

[1] This and the following three problems are not given by Robert of Chester, although the Libri and Arabic versions give the first, second, and fourth problems, and the Boncompagni text the first and second. Scheybl was probably familiar with the other Latin text, the one published by Libri. The geometrical figure in the Libri text is an L-shaped figure, and the same reversed in the Arabic text. In each of these the vertical line represents $\sqrt{200}$.

[2] The geometrical demonstration of this simple algebraic addition is relatively complicated; in algebraic symbolism, $\sqrt{200} - 10 + 20 - \sqrt{200} = 10$, and similarly below.

1   *Secundum aenigma*

Radix ex 200, subtractis 10, diminuta a duplo subtracti, scilicet a 20 excepta
radice ex 200 est triplum subtracti, scilicet 30, praeter duas radices ducentorum.

Dicit aenigma, si a radice numeri 200 subtrahantur 10, id deinde quod relinquitur
5 a 20 subtrahatur, atque ex hoc residuo postea radix numeri 200 auferatur, quod
tandem triplum subtracti, hoc est 30, et duae radices numeri 200 maneant.   Hoc
autem sic probabitur.   Sit linea *a b* radix ducentorum, et sit *b c* sibi aequalis, *b d*
verò 20, atque postea *a e* 10.   Secabo autem de linea *b a* lineam *b f* aequalem *a e*.
Erit igitur tota linea *d f* 30.   Porro ex linea *c d* secabo lineam *c h* aequalem lineae
10 *e b*.   Et quoniam linea *e b* est radix ducentorum exceptis 10, cum linea *c d* sit 20
excepta radice ducentorum, facta subtractione *e b* lineae a linea *c d*, manet linea
h d quae erit 30 exceptis duabus radicibus ducentorum.   Quòd autem haec eadem
linea *h d* praecise tantum sit, probabo.   Linea *f d* est 30 et cum linea *b c* radix sit
ducentorum, linea deinde *a e* lineae *f b*, linea etiam *e b* lineae *c h* aequalis; et
15 aggregatum ergo ex lineis *f b* et *c h* radix ducentorum erit.   Manet autem subtractione facta linea *h d*.   Probatum ergò quod probandum erat.

Sequitur figura.

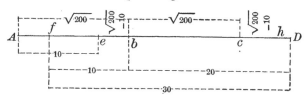

Tertium aenigma

Duae radices alicuius numeri sunt vna sui quadrupli.
20   Hoc satis patet ex eo quòd quadratum est quadruplum ad aliud quadratum,
cuius costa est dupla ad costam alterius.

Quartum aenigma

Centum et census exceptis 20 radicibus adiuncti ad 50 et ad 10 radices exceptis
duobus censibus, sunt 150 exceptis censu et 10 radicibus.
25   Probatio.   Vbi enim subtrahuntur 20 radices, et adduntur 10 radices, idem est
ac si solum 10 radices subtraherentur, et vbi additur census et subtrahuntur duo
census, idem est ac si solum modo vnus census subtraheretur, ex quo patet propositum.

Sequuntur nunc harum regularum exercitii maioris causa quaestiones decem et octo

30   *Prima*

Diuisi 10 in duas partes et multiplicaui vnam partem cum altera, et postea
vnam cum seipsa, et produxit haec multiplicatio partis cum seipsa tantum quantum multiplicatio vnius partis cum altera quater: quae igitur fuerunt partes?

Second problem

The root of 200 less 10, taken from twice that which is subtracted, namely from 20, less the root of 200, is the triple of that which was first subtracted, namely 30, less two roots of 200.

The problem states that when you subtract 10 from the root of the number 200, and this in turn from 20, and from the remainder you take the root of the number 200, then the triple, 30, of the quantity originally subtracted, 10, less [1] two roots of the number 200 remains. This will be proved in the following manner:

Let the line $a\,b$ represent the root of 200, and let $b\,c$ be equal to it. Let $b\,d$ be 20, and further $a\,e$ be 10. Cut off from the line $b\,a$ the line $b\,f$ equal to $a\,e$. Then the whole line $d\,f$ is 30. Further, from the line $c\,d$ cut off the line $c\,h$ equal to the line $e\,b$. Since the line $e\,b$ is the root of 200, less 10, and the line $c\,d$ is 20 less the root of 200, making the subtraction of the line $e\,b$ from the line $c\,d$, you obtain the line $h\,d$ which is then 30 less two roots of 200.

Moreover that this line $h\,d$ is exactly of that length, I will prove. The line $f\,d$ is 30, and since the line $b\,c$ is the root of 200, the line $a\,e$ equals $f\,b$, and also the line $e\,b$ equals $c\,h$. Hence by adding the two lines $f\,b$ and $c\,h$ we shall obtain the root of 200. The subtraction being made there remains the line $h\,d$. That therefore which was to be proved, has been proved.

Third problem

Two roots of any number make one of the quadruple.

This is sufficiently evident from the fact that any square is the quadruple of another square if the side of the first is double the side of the other.

Fourth problem

One hundred and x^2 less 20 x, added to 50 plus 10 x less 2 x^2, gives 150 less x^2 and less also 10 x.

Proof. Where 20 roots are subtracted and 10 roots are added, the result is the same as if only 10 roots were subtracted. Also where a square is added and two squares are subtracted, the result is the same as if only one square were subtracted. From this follows that which was proposed.

There follow eighteen questions for greater practice in these rules [2]

First question

I divided 10 into two parts and multiplied one part by the other, then I multiplied one part by itself. This product of one part by itself gave as much as four times the product of one part by the other. What were the two parts? The answer by the rule is 8 and 2.

---

[1] Not "and," as in the text.

[2] The eighteen problems are all, with minor changes, from Al-Khowarizmi's algebra, but problems 10, 11, and 14, which follow, Robert did not include in his text:

(10) $\dfrac{x}{10-x}+\dfrac{10-x}{x}=2\frac{1}{6}$; (11) $\dfrac{1}{2}\cdot\dfrac{5\,x}{10-x}+5\,x=50$; (14) $\dfrac{x\,(10-x)}{10-2\,x}=5\frac{1}{4}$.

1   Respondetur ex regula: 8 et 2.

Regula. Vnam sectionem ponas rem et alteram 10 praeter rem. Multiplica deinde rem cum 10 praeter rem, et fiunt 10 res praeter censum. Postea quadrupla hoc totum et producuntur 40 res praeter 4 census, et illud aequatur productum ex
5 multiplicationis rei cum re, quod est census. Ex hypothesi ergo quaestionis census aequalis est 40 rebus praeter 4 census. Vnus igitur census aequatur 8 rebus. Sed ille est 64, radix vero numerus 8. Vna igitur sectio est 8, et per consequens reliqua 2. Ista autem quaestio reducitur ad caput vbi census radicibus aequatur.

10   *Quaestio secunda*

Diuisi 10 in duas partes, et multiplicaui, 10 cum seipso, fuitque resultans ex tali multiplicatione aequale vni duarum partium multiplicatae cum seipsa bis et septem nonis multiplicationis vnius: quae igitur fuerunt partes?

Respondetur ex regula: 6 et 4.

15 Regula. Ponatur vna duarum partium res, quae multiplicetur cum seipsa, et fit census. Is census dupletur et addantur septem nonae census et resultant 2 et $\frac{7}{9}$, hoc est viginti quinque nonae, census. Igitur si census esset nouem partes, istae essent quintae pars et quatuor quintae vnius quintae totius, quod esset viginti quinque nonae. Sed ex hypothesi totum multiplicatum vel productum
20 est 100. Accipiatur igitur quinta pars et $\frac{4}{5}$ quintae partis ex numero 100, erunt autem haec 36. Atque tantus est census, cuius radix scilicet 6 vna diuisionis pars in proposita quaestione. Reducitur aut haec quaestio ad caput in quo census numero aequatur.

                    Quaestio tertia

25 Diuisi 10 in duas partes, et diuisi vnam partem in aliam et exiuerunt 4. Quaeritur, quae sint partes.

Respondetur ex regula: 8 et 2.

Regula. Sit vna pars res, et altera 10 praeter rem. Si igitur ex diuisione 10 praeter rem in rem exierint 4. Igitur ex multiplicatione 4 cum re resultabunt 10
30 praeter rem. Quatuor igitur res sunt 10 praeter rem, atque ita quinque res decem praecise. Ergò vna res 2. Reducitur autem haec quaestio ad illud caput vbi res numero aequatur.

                    Quaestio quarta

Multiplicaui tertiam partem census et denarium vnum cum quarta parte
35 census et cum denario vno, et prouenerunt 63: quantus igitur est census?

Respondetur ex regula: 24.

Regula. Multiplica tertiam partem cum quarta, et producetur pars duodecima census; deinde multiplica vnum denarium cum $\frac{1}{4}$ et producetur $\frac{1}{4}$ rei; multiplica etiam tertiam cum denario et producetur $\frac{1}{3}$ rei, atque tandem denarium
40 cum denario, et producetur denarius. Est autem totum multiplicationis productum $\frac{1}{12}$ census, quarta rei et tertia rei atque denarius, aequantes 63 denarios. Demas igitur primò denarium vtrique, casum deinde hunc, vt quidem semper

Rule. You would let one part equal x and the other $10 - x$; then multiply x by $10 - x$, giving $10 x - x^2$. Afterwards take the quadruple of the total, which gives $40 x - 4 x^2$. This equals the product of x multiplied by x, which is x^2. Now under the supposition of our question, x^2 is equal to $40 x - 4 x^2$. Hence x^2 equals $8 x$, whence x^2 is 64, and the root of it is the number 8. One of the parts of 10 is then 8, and consequently the remaining part is 2. This problem, then, is reduced to the chapter wherein a square is equal to roots.

Second question

I divided 10 into two parts, and multiplied 10 by itself. The result of this multiplication was equal to the product of one of these parts multiplied by itself and by two and seven-ninths. What are the parts ? The answer in accordance with the rule is 6 and 4.

Rule. Let x represent one of the two parts, which multiplied by itself becomes x^2. Double this square and add to it seven-ninths of one square, giving $2\frac{7}{9} x^2$ or $\frac{25}{9} x^2$. Hence if the whole square consists of nine parts (ninths) that will be the fifth part and four-fifths of one-fifth of the whole, which is twenty-five ninths. But under our supposition the product is equal to 100. Therefore take one-fifth, and $\frac{4}{5}$ of one-fifth, of the number 100, namely 36, which is the value of x^2. The root of this, namely 6, is one part of 10 in the proposed problem. This question then is referred back to the chapter in which a square is equal to a number.

Third question

I divided 10 into two parts in such a way that the one part divided by the other equaled 4. The question is, what are the parts? The answer in accordance with the rule is 8 and 2.

Rule. Let x represent one part, and $10 - x$ the other. Now if by the division of $10 - x$ by x there results 4, it follows that by the multiplication of 4 by x you will obtain $10 - x$; $4 x$, then, equals $10 - x$, and hence 10 precisely equals $5 x$. The value of x is 2. This question is referred back to that chapter in which a root is equal to a number.

Fourth question

I multiplied one-third x^2 plus one unit by one-fourth x^2 plus one unit and the result was 63. What is the value of x^2? The answer in accordance with the rule is 24.

Rule. Multiply one-third (x) by one-fourth (x), and one-twelfth x^2 will be the result; then multiply one unit by $\frac{1}{4}$ (x), and $\frac{1}{4} x$ will be the result; again multiply $\frac{1}{3}$ (x) by one unit, giving $\frac{1}{3} x$, and finally a unit by a unit, giving a unit. The sum total of this multiplication is $\frac{1}{12} x^2 + \frac{1}{4} x + \frac{1}{3} x + 1$ which equals 63 units. Take one unit from both sides, and then, according to the rule that it is always necessary to reduce to an integral square, the whole

1 necesse, ad censum integrum reducas, totum scilicet, vel singula, cum 12 multiplicando, et venient 1 census et 7 res aequales 744 denariis. Per primam igitur
regulam dimidia radices, et medietatem cum seipsa multiplica et producentur
$12\frac{1}{4}$; quibus additis ad 744 resultabunt $756\frac{1}{4}$. Hinc elice radicem quadratam,
5 veniunt $27\frac{1}{2}$, atque ab his medietatem radicum 3 scilicet et $\frac{1}{2}$ subtrahe et relinquuntur 24, quantitas census. Haec autem quaestio ad caput in quo census et
res numero aequales sunt, reducitur.

Quaestio quinta

Diuisi 10 in duo, et multiplicaui vtrumque cum seipso, et aggregatum quadra-
10 torum fuit 58: quae igitur sunt partes?
Respondetur ex regula: 7 et 3.
Regula. Multiplica 10 praeter rem cum seipsa, et producentur 100 den. et
census, exceptis 20 rebus; deinde multiplica rem cum seipsa, et producetur census.
Habes igitur 100 den. et 2 census, exceptis 20 rebus, quae omnia aequantur 58.
15 Restaurando igitur 20 res diminutas, patet quòd 100 denarii et 2 census valent
58 den. et 20 res, atque reducendo ad vnum censum, veniunt 50 den. et 1 census
aequales 29 den. et 10 rebus. Postea vero subtrahendo 29 a 50 relinquuntur 21
den. et vnus census aequales 10 rebus. Operare igitur per caput secundum, et patet
quòd altera pars 7 et altera 3 sint. Haec autem quaestio reducitur ad caput in
20 quo census et numeri rebus aequantur.

Questio sexta

Multiplicaui tertiam census cum quarta eiusdem, et prouenerunt census et 24
denarii: quantus igitur est census?
Respondetur ex regula: 24.
25 Regula. Quia satis nosti, quòd ex multiplicatione tertiae partis rei cum quarta
parte rei, duodecima proueniat pars census, quae in hoc casu vni rei et 24 denariis
aequalis est. Integra igitur censum, multiplicando totum cum 12, et perficias
quod census vnus 12 rebus et 288 denariis aequetur. Age igitur, multiplicando
medietatem radicum cum seipsa, et addendo productum seu quadratum ad 288,
30 et prouenient 324, cuius radix est 18, quae addita ad medietatem radicum, veniunt
24, census propositae quaestionis. Reducitur autem haec quaestio ad caput in
quo radices et numeri substantias coaequant.

Quaestio septima

Diuisi 10 in duo, vt multiplicatio vnius cum altero producat 24: quae igitur
35 sint partes, quaeritur.
Respondetur ex regula: 6 et 4.
Regula. Scias vnam partem esse rem, et alteram 10 praeter rem. Multiplica
igitur vnam cum altera, et producuntur 10 res praeter censum, quae valent 24
denarios. Restaurando igitur dic, quòd 10 res valeant 24 den. et vnum censum.
40 Age nunc per caput quo census et numeri rebus aequantur, et patebit quòd vna
pars diuisionis 6 et altera 4 sint.

or each part is now multiplied by 12; you will obtain $x^2 + 7x$ equal to 744 units. By the first rule you take one-half of the roots and multiply the half by itself, obtaining $12\frac{1}{4}$, which being added to 744 will make a total of $756\frac{1}{4}$. Take the square root of this and you obtain $27\frac{1}{2}$. Now from $27\frac{1}{2}$ subtract the half of the roots, namely 3 and $\frac{1}{2}$, and 24 remains as the value of the square. This question is referred back to the chapter in which squares and roots are equal to a number.

Fifth question

I divided 10 into two parts and I multiplied each part by itself, and the sum of the squares was 58. What are the two parts? The answer in accordance with the rule is 7 and 3.

Rule. Multiply $10 - x$ by itself, and you obtain 100 units $+ x^2 - 20x$; then multiply x by itself, and x^2 is obtained. You have then 100 units $+ 2x^2 - 20x$, a total equal to 58. By restoration then of the negative $20x$, it follows that 100 units $+ 2x^2$ equal 58 units $+ 20x$, and reducing this to one square, 50 units $+ x^2$ are found equal to 29 units $+ 10x$. Accordingly, by subtracting 29 from 50, 21 units $+ x^2$ remain equal to $10x$. Proceed therefore according to the second chapter, and it becomes clear that the parts are 7 and 3. This problem is referred back to the chapter in which squares and numbers are equal to roots.

Sixth question

I multiplied $\frac{1}{3}x$ by $\frac{1}{4}$ of it and obtained $x + 2$ units. What is the value of x? The answer in accordance with the rule is 24.

Rule. Since you well know that the product of $\frac{1}{3}x$ by $\frac{1}{4}x$, is $\frac{1}{12}x^2$, it follows that $\frac{1}{12}x^2$, in this instance, equals $x + 24$ units. Make the square whole by multiplying all by 12, and you find that $x^2 + 12x$ is equal to 288 units. Treat this (equation), then, by multiplying the half of the roots by itself and adding the product, or square, to 288. You obtain 324, whose root is 18; this added to the half of the roots gives 24, the square which was sought in the proposed question. This problem is referred back to the chapter in which roots and numbers are equal to squares.

Seventh question

I divided 10 into two parts in such a way that the product of one by the other is 24. The question is, what are the parts? The answer in accordance with the rule is 6 and 4.

Rule. You know that you can let the one part equal x, and the other $10 - x$. Then multiply the one by the other and you obtain $10x - x^2$, which equals 24 units. Now by restoration say that $10x$ amounts to 24 units $+ x^2$. Treat this then according to the chapter in which squares and numbers are equal to roots, and it will be clear that one part is 6 and the other 4.

1   *Quaestio octaua*

Diuisi 10 in duo, atque vtroque multiplicata cum seipso, subtraxi minus de ma-
iori et manserunt 40.   Quaeritur de duobus.
                    Respondetur ex regula : 7 et 3.
5 Regula.   Multiplica rem cum re, et proueniet census.   Deinde multiplica
etiam 10 praeter rem cum 10 praeter rem, et prouenient 100 dena. et census,
exceptis 20 rebus.   Subtrahe igitur censum a 100 et censu exceptis 20 rebus, et
manent 100 exceptis 20 rebus quae aequantur 40 dena.   Restaurando igitur dic,
quòd 100 denarii aequentur 40 denariis et 20 rebus.   Subtrahendo deinde 40 a 100
10 patet quòd 60 denarii aequantur 20 rebus; tres igitur denarii rei vni; estque
ternarius vna diuisionis pars, quare 7 altera.

                               Quaestio nona

Diuisi 10 in duo et vtramque multiplicaui cum seipso, adiunxi deinde producta
simul et insuper addidi differentiam horum duorum antequam multiplicarentur
15 cum seipsis, et prouenerunt 54.   Quaeritur.
                    Respondetur ex regula : 6 et 4.
Regula.   Multiplica rem cum re, deinde etiam 10 praeter rem cum 10 praeter
rem, et adde simul producta, et veniunt 100 denarii et 2 census, exceptis 20 rebus.
Cum igitur superfluum inter partes, vel partium differentia, sit 10 praeter 2 res,
20 adiecto hoc superfluo, erit totum 110 dena. et 2 census exceptis 22 rebus, quae
omnia aequantur 54 denariis.   Dic igitur, res integrando, quòd 110 den. et 2
census aequentur 54 denariis et 22 rebus.   Reducendo deinde ad vnum censum,
dic, quòd 55 denarii et 1 census aequentur 27 denariis et 11 rebus; subtrahendo
verò 27 a 55, dic quòd 28 denarii et vnus census aequentur 11 rebus.   Age igitur
25 per caput quo census et numeri rebus aequantur, et patebit quòd vna pars diuisionis 6 et altera 4 erunt.

                              Quaestio decima

Diuisi 10 in duo et vtrunque diuisi in alterum, et aggregatum ex diuisionibus,
id est exeuntium, fuit 2 et vna sexta.   Quaeritur.
30   Respondetur ex regula : 4 et 6.
Regula.   Aggregatum ex multiplicatione vtriusque cum seipso, aequum est
multiplicationi vnius cum altero et producti cum aggregato ex diuisionibus
vtriusque in alterum quod in hoc casu est 2 et vna sexta.   Ideo multiplica 10
praeter rem cum seipso, et re cum re, et prouenient 100 et 2 census, exceptis 20
35 rebus, et hoc totum aequatur multiplicationi rei cum 10 praeter rem, et producti
cum 2 et vna sexta.   Sed multiplicatio rei cum 10 praeter rem producit 10 res
praeter censum, quibus cum 2 et $\frac{1}{6}$ multiplicatis 21 res et $\frac{2}{3}$ rei praeter $2\frac{1}{6}$ census
resultabunt, quae aequantur 100 dena. et 2 censibus exceptis 20 rebus.   Re-
staurando igitur census et res, veniunt 41 res et $\frac{2}{3}$ rei aequales 100 denariis et 4
40 censibus cum sexta parte census.   Reduc igitur totum ad vnum censum sic.
Consideretur censum esse 6, et erunt 4 census et sexta, 25.   Atque huius vnus
census, 6 scilicet, est vna quinta et quinta quintae.   Totius igitur quod habes

Eighth question

I divided 10 into two parts, and each being multiplied by itself, I subtracted the smaller from the larger, and 40 remained. The question is as to the parts? The answer in accordance with the rule is 7 and 3.

Rule. Multiply x by x, and you obtain x^2. Then multiply also $10 - x$ by $10 - x$ and you obtain 100 units $+ x^2 - 20 x$. Subtract, therefore, x^2 from $100 + x^2 - 20 x$, and you have $100 - 20 x$, which equals 40 units. By restoration say, then, that 100 units are equal to 40 units and $20 x$. Then by subtracting 40 from 100 it is plain that 60 units are equal to $20 x$, and hence 3 units to one x. Three is one part; hence 7 is the other.

Ninth question

I divided 10 into two parts, and multiplied each part by itself; then I added these products together, and the difference between these two, before each was multiplied by itself, and the result was 54. The question is stated. The answer in accordance with the rule is 6 and 4.

Rule. Multiply x by x, and also $10 - x$ by $10 - x$; add the products and you obtain 100 units $+ 2 x^2 - 20 x$. Since the excess of one part over the other, or the difference of the parts, is $10 - 2 x$, when this excess is added we shall have 110 units $+ 2 x^2 - 22 x$ as the sum total, which equals 54 units. Say therefore, by adding the $22 x$, that 110 units $+ 2 x^2$ equals 54 units $+ 22 x$. Then by reduction to one square, you say that 55 units $+ x^2$ equal 27 units and $11 x$. By subtracting 27 from 55, you say that $x^2 + 28$ units equals $11 x$. Treat this by the chapter in which squares and numbers equal roots, and it will be plain that one part is 6, and the other 4.

Tenth question

I divided ten into two parts, and I divided each of these by the other; the sum of the two quotients, that is to say, the result, is two and one-sixth. The question is stated. The answer in accordance with the rule is 4 and 6.

Rule. The sum of the products of each (of the parts) multiplied by itself is equal to the product of the one by the other when this product is multiplied by the sum of the quotients of each of the divisions, which in this case is $2\frac{1}{6}$. Hence you multiply $10 - x$ by itself, and x by x, obtaining $100 + 2 x^2 - 20 x$. This total is equal to the product of x by $10 - x$, multiplied by $2\frac{1}{6}$. But the product of x by $10 - x$ gives $10 x - x^2$, which being multiplied by $2\frac{1}{6}$ there results $21 x + \frac{2}{3} x - 2\frac{1}{6} x^2$ equal to 100 units $+ 2 x^2 - 20 x$. By restoration, then, of the squares and the roots, $41 x + \frac{2}{3} x$ are obtained equal to 100 units $+ 4 x^2 + \frac{1}{6} x^2$. Reduce the whole then to one square in the following manner: If a square is supposed to be 6, then $4\frac{1}{6}$ squares would be 25; of this, one square, namely 6, is $\frac{1}{5}$ and $\frac{1}{5}$ of $\frac{1}{5}$. Take therefore of everything which you have $\frac{1}{5}$ and $\frac{1}{5}$ of $\frac{1}{5}$, and it will be plain that $10 x$ is equal to $x^2 + 24$ units. Proceed then by that chap-

1 accipe quintam partem et quintam quintae, et patebit, quòd 10 res vni censui et
24 denariis aequentur.   Age igitur per caput, quo census et numeri rebus aequantur, multiplicando medietatem radicum cum seipsa, et producuntur 25; a quibus
subtrahe 24 et manet vnitas, cuius radix est vnitas.   Hanc radicem subtrahe a
5 medietate radicum, et manent 4, quae sunt vna diuisionis pars.   Et nota, quòd
cum illud quod resultat ex diuisione primae partis in secundam alicuius totius,
multiplicetur cum illo quod resultat ex diuisione secundae partis in primam, illud
quod prouenit semper idem sit.

Quaestio vndecima

10   Diuisi 10 in duas partes et multiplicaui vnam illarum cum 5 et productum diuisi
in reliquam partem, et exeuntis medietatem addidi ad productum ex multiplicatione primae partis cum 5, et totum aggregatum fuit 50.   Quaeritur.
Respondetur ex regula : 8 et 2.
Regula.   Ex 10 accipias rem; hanc multiplicabis cum 5, et fient 5 res.   De-
15 beres diuidere 5 res in 10 praeter rem, et addere medietatem exeuntis ad 5 res.
Sed hoc idem est, ac si diuideres medietatem 5 rerum in 10 praeter rem, et adderes
exeuntem totum ad 5 [res].   Vtraque enim operatio producitur 50.   Si ergo diuidas 2
res et semissem in 10 praeter rem, exeunt 50 praeter 5 res, eo quod addito producto
ad 5 res, prouenirent 50.   Cum igitur constet quod multiplicato illo quod prouenit
20 ex diuisione cum diuisore redeat census tuus, qui est 2 res et semis.   Multiplica
igitur 10 praeter rem cum 50 exceptis 5 rebus, prouenient 500 et 5 census exceptis
100 rebus, quae omnia aequantur duabus rebus et semissi.   Reduc igitur totum ad
vnum censum accipiendo quintam partem totius, et patebit, quòd 100 et census,
exceptis 20 rebus aequentur medietati rei.   Restaurando igitur, dic quod 100 et
25 census aequentur 20 rebus et medietati rei.   Age igitur per caput quo census et
numeri rebus aequantur, multiplicando medietatem rerum cum se, et prouenient
105 et $\frac{1}{16}$; a quibus subtractis 100 manent $5\frac{1}{16}$, cuius radix est $2\frac{1}{4}$; quibus subtractis a medietate radicum, et manent 8 vna diuisionis pars.

Quaestio duodecima

30   Diuisi 10 in duos partes, et multiplicatio vnius partis cum seipsa produxit
numerum continentem alteram partes octogesies semel.   Quaeritur de partibus.
Respondetur ex regula : 9 et 1.
Regula.   Multiplica 10 praeter rem cum se, et fient 100 et census, praeter 20
res quae aequantur 81 rebus.   Restaurando igitur dic, quòd 100 et census aequen-
35 tur 101 rebus.   Age nunc per caput quo census et numeri rebus, aequantur, et
veniet tandem vnitas, vna diuisionis pars.

Quaestio decima tertia

Duo sunt census, quorum maior excedit minorem in duobus, diuisi autem
maiorem in minorem, et exibat medietas maioris.   Quaeritur.
40   Respondetur ex regula : 2 census minor et 4 maior.
Regula.   Pone rem pro censu, et dic, quia res minor diuidens maiorem producit

ter in which a square and numbers are equal to roots. Multiplying one-half of the roots by itself you have 25; from this subtract 24 and there remains one, of which the root is one. Subtract this root from the half of the number of the roots, and four remains as the value of one part. Now note that when the quotient obtained by dividing the first part by the second part of any whole is multiplied by the quotient of the second part by the first, that which is obtained is always the same.

Eleventh question

I divided ten into two parts and I multiplied one of these by five, and the product I divided by the other part; one-half of this result I added to the product of the first part multiplied by 5, and the sum total was fifty. The question is stated. The answer in accordance with the rule is 8 and 2.

Rule. You may take x as one part of 5; this you will multiply by 10, giving $5x$. You should divide $5x$ by $10 - x$, and add $\frac{1}{2}$ of the quotient to $5x$. But this is the same as if you should divide $\frac{1}{2}$ of $5x$ by $10 - x$ and add the total result to $5x$; either operation gives 50. If therefore you divide $2\frac{1}{2}x$ by $10 - x$, $50 - 5x$ is obtained, since when $5x$ was added to the quotient, the sum was given as 50. Moreover it should be evident that the product of the result of any division multiplied by the divisor gives your quantity (the dividend), which is $2\frac{1}{2}x$. Therefore multiply $10 - x$ by $50 - 5x$, obtaining $500 + 5x^2 - 100x$, all of which is equal to $2\frac{1}{2}x$. Reduce the whole then to one square by taking the fifth part of the whole, and it will be clear that $100 + x^2 - 20x$ equals $\frac{1}{2}x$. By restoration then say that $100 + x^2$ equals $20x$ and $\frac{1}{2}x$. Operate then by the chapter in which squares and numbers are equal to roots. Multiplying one-half of the roots by itself, 105 and $\frac{1}{16}$ is obtained; from this you subtract 100, leaving $5\frac{1}{16}$, of which the root is $2\frac{1}{4}$, and this being subtracted from one-half of the roots 8 remains as the value of one part.

Twelfth question

I divided 10 into two parts, and the product of one of these parts by itself contained the other part 81 times. The question is as to the parts. The answer in accordance with the rule is 9 and 1.

Rule. Multiply $10 - x$ by itself, giving $100 + x^2 - 20x$, which is equal to $81x$. Then by restoration, say that $100 + x^2$ equals to $101x$. Operate now by the chapter in which squares and numbers are equal to roots, and unity will finally appear as the value of one part.

Thirteenth question

There are two quantities of which the greater exceeds the less by two. I divided the greater by the less and the quotient was one-half the greater quantity. The question is stated. The answer in accordance with the rule is 2, for the smaller quantity, and 4 for the larger.

1 medietatem rei maioris, ideo econtra res minor multiplicata cum medietate rei maioris producit rem maiorem, et duo multiplicata cum medietate rei maioris, producit rem maiorem. Binarius igitur est res minor, et quaternarius maior.

Quaestio decima quarta

5 Diuisi 10 in duas partes, et multiplicaui vnam partem cum altera, et productum diuisi in differentiam inter partes, et resultarunt $5\frac{1}{4}$. Quaeritur, quae sint partes. Respondetur ex regula: 3 et 7.

Regula. Multiplica rem cum 10 praeter rem, et fient 10 res excepto censu; deinde diuide 10 res excepto censu in 10 exceptis 2 rebus, quae sunt differentia inter
10 partes, et exeunt $5\frac{1}{4}$. Si igitur econtrà multiplicaueris $5\frac{1}{4}$ cum 10, exceptis 2 rebus, prouenient 10 res excepto censu. Multiplica igitur $5\frac{1}{4}$ cum 10, exceptis 2 rebus, et producuntur 52 den. et semis praeter 10 res et semissem. Atque haec omnia aequantur 10 rebus, excepto censu. Dic igitur, restaurando res et denarios, quòd $20\frac{1}{2}$ res aequentur $52\frac{1}{2}$ denariis et vni censui. Age igitur per caput quo census
15 et numeri rebus aequantur, multiplicando medietatem radicum in se, et prouenient $105\frac{1}{16}$ et quae sequuntur et caet.

Quaestio decima quinta

Quatuor radices census multiplicatae cum quinque radicibus eiusdem census, producunt duplum census et 36. Quaeritur de censu.
20   Respondetur ex regula: 2.

Regula. Multiplica 4 res cum 5 rebus, et fiunt 20 census qui aequantur 2 censibus et 36 denariis. Diuide ergo 36 in 18 et exeunt 2. Atque tantus est census, quod examinari poterit.

Quaestio decima sexta

25 Subtraxi a censu eius vnam tertiam et tres denarios, multiplicaui deinde residuum cum seipso, restituit haec multiplicatio ipsum censum. Quantus igitur census sit, quaeritur.
Respondetur ex regula: 9.

Regula. Subtracta tertia et tribus a tribus tertiis rei, manent $\frac{2}{3}$ rei praeter 3 dena.
30 quae sunt radix census. Multiplica igitur $\frac{2}{3}$ rei praeter 3 den. cum se, et producentur $\frac{4}{9}$ census et 9 den. praeter 4 res, et illud aequatur radici. Ergo $\frac{4}{9}$ census et 9 denarii valent 5 res. Reducas $\frac{4}{9}$ ad vnum censum, eundem, denarios etiam et res cum duobus et quarto multiplicando, et inuenies, quòd census et $20\frac{1}{4}$ denarii aequantur 11 rebus et $\frac{1}{4}$. Age igitur per caput quo census et numeri rebus aequan-
35 tur. Accipiendo medietatem radicum quae est $5\frac{5}{8}$ et multiplicando eam cum seipsa, et fiunt $31\frac{41}{64}$, de quibus subtrahe $20\frac{1}{4}$ et manent $11\frac{25}{64}$, cuius radix est $3\frac{3}{8}$, quam ad medietatem radicum adde, quia non per subtractionem non deunies ad intentum, et veniunt 9, census qui quaerebatur.

Rule. Let x represent the one quantity, and say, since the lesser quantity divided by the greater gives one-half of the greater, that consequently the lesser multiplied by one-half of the greater gives the greater quantity. But two times one-half of the greater quantity gives also the greater quantity. Therefore 2 is the value of the lesser quantity, and 4 is the greater.

Fourteenth question

I divided 10 into two parts, and I multiplied one by the other and divided the product by the difference between the two, obtaining $5\frac{1}{4}$ as the result. The question is, what are the parts? The answer is 3 and 7.

Rule. Multiply x by $10 - x$, giving $10x - x^2$; then divide $10x - x^2$ by $10 - 2x$, which is the difference between the parts, and $5\frac{1}{4}$ is obtained. Now, on the other hand, if you multiply $5\frac{1}{4}$ by $10 - 2x$, you will obtain $10x - x^2$. Hence multiply $5\frac{1}{4}$ by $10 - 2x$, which gives $52\frac{1}{2}$ units $- 10\frac{1}{2}x$, all of which is equal to $10x - x^2$. Observe, then, that by restoring to the $10x$ and to the units (the quantities, x^2 and $10\frac{1}{2}x$ respectively, which are subtracted from them) $20\frac{1}{2}x$ is equal to $52\frac{1}{2}$ units $+ x^2$. Operate therefore by the chapter in which squares and numbers are equal to roots, multiplying the half of the roots by itself, and there will result $105\frac{1}{16}$, etc.

Fifteenth question

Four roots of a square multiplied by five roots of the same square give double the square and 36. The question is as to the square. The answer in accordance with the rule is 2.

Rule. Multiply $4x$ by $5x$, giving $20x^2$, which equals $2x^2 + 36$ units. Hence divide 36 by 18, giving 2 as the result. And this amount is the square, which may be tested.

Sixteenth question

I subtracted from a quantity one-third of it and three units, then I multiplied the remainder by itself, restoring the quantity itself by this multiplication. The question is, how great is the quantity? The answer in accordance with the rule is 9.

Rule. Subtracting $\frac{1}{3}x + 3$ (units) from $\frac{3}{3}x$, there remain $\frac{2}{3}x - 3$ units, which is the root of the quantity (x). Therefore multiply $\frac{2}{3}x - 3$ units by itself, giving $\frac{4}{9}x^2 + 9$ units $- 4x$, and that is equal to the root (x). Hence $\frac{4}{9}x^2 + 9$ units equals $5x$. You reduce the $\frac{4}{9}$ to a whole square by multiplying it, and also the units and the $5x$, by $2\frac{1}{4}$, and you will find that $x^2 + 20\frac{1}{4}$ units is equal to $11\frac{1}{4}x$. Operate then by the chapter in which squares and numbers are equal to roots. Taking the half of the roots, $5\frac{5}{8}$, and multiplying it by itself, you have $31\frac{41}{36}$: from this subtract $20\frac{1}{4}$, there remains $11\frac{25}{64}$, of which the root is $3\frac{3}{8}$. Add this to the half of the roots, since by subtraction you will not arrive at the desired result, and 9 appears as the quantity which you seek.

1 *Quaestio decima septima*

Diuisi drachmam et semissem inter homines et partem hominis, et contigit homini duplum eius quod parti. Quanta igitur fuerit pars, quaeritur.

Respondetur ex regula : $\frac{1}{2}$.

5 Regula. Idem est homo et pars, ac si diceres, vnum et res. Diuidatur ergo drachma et semis in vnum et rem, et venient 2 res. Multiplica deinde 2 res cum drachma et re et fient 2 census et 2 res, quae aequantur drachmae et semissi. Reducendo igitur ad vnum censum, dic quòd census et res, aequentur $\frac{3}{4}$ drachmae. Age igitur per caput, quo census et res numero coaequantur, multiplicando me-
10 dietatem rei in seipsa, et fit quarta, quae addita ad $\frac{3}{4}$ facit vnum, cuius radix est vnum, a qua subtrahe medietatem rei et manet medietas, pars quae quaeritur.

Quaestio decima octaua

Diuisi drachmam inter homines et prouenit simul res, deinde addidi eis hominem et postea diuisi drachmam inter eos, et cuilibet contigit minus quod prius sexta
15 parte drachmae. Quot igitur fuerint homines, quaeritur.

Respondetur ex regula : 2.

Regula. Huius quaestio consideratio est vt multiplices homines primos cum diminuto inter diuisiones; deinde multiplices aggregatum cum hominibus primis et cum homine addito, et proueniet census tuus. Scias autem, quòd hoc non sit
20 vniuersaliter verum vt credo, sexta census et sexta radicis, quae aequantur drach-mae. Dic ergo, res integrando, quòd census et res aequentur 6 drachmis. Age igitur per caput quo census et res numeris coaequentur. Multiplicando medie-tatem radicum cum seipsa, et fit quarta, quam adde ad 6 drachmas et veniunt $6\frac{1}{4}$. Inde radix quadrata erunt 2 et semis a qua subtracta medietate radicum, manent
25 2, qui est numerus hominum.

Sequitur vltimo de rebus venalibus

Sunt autem conuentiones negociationum quae fiunt in venditione, emptione, permutatione, et caeteris rebus, secundum duos modos.

Primus est modus, vt si dicatur, decem res venditae sunt 6 drachmis, quot
30 igitur veniunt 4 drachmis ?

Secundus modus est, vt si dicatur, decem res venditae sunt 6 drachmis, quantum igitur est precium 4 rerum ?

In primo casu, 10 res est numerus appreciati secundum positionem, et 6 drach-mae est praecium secundum positionem; quaestio quot, est numerus ignotus
35 appreciati secundum quaerentem, et 4 res precium secundum quaerentem. In secundo casu, precium et appreciatum secundum positionem, sunt vt prius, et quaestio et precium secundum quaerentem et 4 res est appreciatum secundum quaerentem. Vnde precium secundum positionem dicitur opponi appreciato secundum quaerentem et appreciatum secundum positionem dicitur opponi
40 precio secundum quaerentem. Multiplica igitur inter se, et productum diuide in tertium modum et exhibit quartus ignotus per regulas quatuor proportionalium quantitatum.

Finis annotationum pro declaratione regularum Algebrae.

Seventeenth question

I divided a drachma and one-half between a man and a part of a man, and to the man there fell the double of that which fell to the part (of a man). The question is, how large was the part? The answer is $\frac{1}{2}$.

Rule. A man and part of a man is the same as $1 + x$. Hence $1\frac{1}{2}$ is divided by $1 + x$, giving $2x$. Then multiply $2x$ by $1 + x$, giving $2x^2 + 2x$, which is equal to $1\frac{1}{2}$. Therefore by reduction to one square, say that $x^2 + x$ is equal to $\frac{3}{4}$ of a unit. Operate now by the chapter in which squares and roots equal number, multiplying one-half of the roots by itself, obtaining $\frac{1}{4}$; this added to $\frac{3}{4}$ makes 1, of which the root is 1; from this subtract the half of the number of the roots, giving $\frac{1}{2}$, the value of the part that is sought.

Eighteenth question

I divided a drachma among some men, and each one obtained an unknown amount (x); I then added one man to the group and again I divided a drachma among them; to each man there now fell $\frac{1}{6}$ drachma less than before. The question is, how many men were there? The answer is 2.

Rule. In considering this problem you multiply the first number of men by the decrease; then you multiply the product by the number of men $+ 1$, and the quantity will be obtained. However you should note that this is not a general rule; you have $\frac{1}{6}x^2 + \frac{1}{6}x$, which is equal to 1. Hence by completing the quantity you obtain $x^2 + x$ equal to 6 units. Operate therefore by the chapter in which squares and roots equal numbers. Multiplying one-half of the roots by itself, $\frac{1}{4}$ is obtained. Add this to the 6 units, giving $6\frac{1}{4}$; then the square root will be $2\frac{1}{2}$. From this one-half of the roots is subtracted, leaving 2, which is the number of men.

The last section, on commercial transactions

There are certain customs of business which hold in buying, selling, exchange, and the like, according to two methods. The first method is illustrated: 10 things are sold for 6 drachmas, how many are sold for 4? The second: 10 things are sold for 6 drachmas, what is the price of 4?

In the first case 10 things is the number priced, according to that which is given, and 6 drachmas is the price, as given; the question, how many, represents the unknown number of things, according to the question, and 4 drachmas the price, according to the question. In the second case, the price and the quantity, as given, are the same as before; the question represents the price, according to the problem, and 4 is the corresponding quantity. Whence the price given is said to be in opposition to the price sought, and similarly the quantity given to that sought. Multiply therefore among themselves, and divide the product by the third kind, and the fourth unknown will appear by the rules of four proportionals.

End of the annotations to explain the rules of algebra.

LATIN GLOSSARY

ABBREVIATIONS

adj.,	adjective.	neut.,	neuter noun.
adv.,	adverb.	num.,	numeral.
comp.,	comparative.	prep.,	preposition.
conj.,	conjunction.	pron.,	pronoun.
demonstr.,	demonstrative.	R,	characterizing a usage of the *Regule*.
f.,	feminine noun.	S,	characterizing a usage peculiar to Scheybl
indef.,	indefinite.		as opposed to Robert of Chester.
indecl.,	indeclinable.	subst.,	substantive.
m.	masculine noun.	super.,	superlative.
n.,	note.	v.,	verb.

The first appearance in the original text of each Latin word listed is recorded by reference to page and line. In a few instances other references are added to indicate differences in meaning. In cases of variation in spelling the catchword presents the form probably used by Robert of Chester; the current form is generally added in parentheses. The limits of the volume preclude a complete study of the Latinity.

abicio, v., *disregard* (76, 2); with ex, *subtract from* (108, 9).

ablatio, f., S, *subtraction* (134, 14).

absque, prep., *minus* (96, 4, n); as adjective, S, *negative* (116, 15).

abstraho, v., with ex, *subtract* (120, 3, n).

accipio, v., *take* (70, 32); *extract* root (72, 1).

addendus (addo), adj., *to be added, positive* (90, 13).

addicio (-tio), f., *addition* (74, 22, n).

additum, neut., S, *addition* (128, 1).

addo, v., *add* (110, 27); with ad, S, *add* (128, 10); with super, *add to* (74, 2, n).

adequo (adaequo), v., *equal* (106, 18, n).

adhibeo, v., *add* to (118, 19).

adicio (adiicio, S), v., *add* to (72, 1); with super, *add to* (72, 1, n).

adiectio, f., *addition* (74, 22).

adiectiuus, adj., *added, positive* (90, 13, n.).

adiectus, adj., *added, positive* (96, 23, n.).

adimpleo, v., *fill up, complete* (74, 21).

aditus, m., *approach* (88, 25).

adiungo, v., *add* (70, 30).

adnullo, v., *reduce to nothing* (74, 29, n).

ae-, see also e-.

aenigma, neut., S, *problem* (142, 17).

aequalitas, f., S, *the equal* (80, 12).

aequatio, f., S, *equation, equality* (66, 2).

aequipolleo, v., S, *equal* (128, 6).

aequus, adj., S, *equal* (150, 31.)

aggregatum, m., S, *sum* (72, 20).

ago, v., *operate* (70, 11).

algaurizim, Arabic, *Al-Khowarizmi* (66, 9, n).

algebra, f., S, *algebra* (66, 1).

aliqui, aliqua, aliquod, indef. pron. adj., *some, any* (74, 21).

aliquot, indef. indecl. num., *some, a number* (90, 9).

almucabola, f., S. A transliteration of part of the Arabic title *al-jebr w'al-muqabala*, for algebra (66, 1).

almuthemen, Arabic, *quantity desired* (120, 24, n).

almuzarar, Arabic, *unit of measure, quantity* (120, 22, n).

alszarar, Arabic, *price per unit* (120, 23, n).

althemen, Arabic, *amount, payment* (120, 24, n).

amplector, v., *embrace, surround, include* (90, 23, n).

angularis, adj., S, *corner* (130, 10).

angulus, m., *an angle, corner* (78, 11).

annotatio, f., S, *annotation* (156, 43).

appareo, v., S, *appear, be evident* (84, 8).

applico, v., *apply, place* (78, 4).

appono, v., *place by, set in apposition* (96, 23, n).

appreciatum, neut., S, *that priced* (156, 36).

arbitror, *consider, judge* (122, 7).

area, f., *area, figure* (78, 4).

arithmeticus, adj., S, *arithmetical* (66, 7).

ars, f., *type* (70, 28).

articulus, m., S, *multiple of ten* (136, 15).

assigno, v., *mark out, indicate* (104, 6).

assimilo, v., R, *make like, be equal* (68, 12, n and 126, 2).

assumo, v., *take, extract* square root (72, 1, n).

attineo, v., *pertain* (120, 21).

auctus, v., S, *positive* (138, 17).

aufero, v., with ex, *take away* (84, 22 ; 110, 21).

augeo, v., S, *increase* (130, 19).

binarium, neut., *two* (98, 1).

binarius, adj., (consisting of) *two* (82, 16).

bis, adv., *twice* (98, 16).

breuis, adj., S, *short* (82, 23).

cado, v., S, *fall (upon)* (132, 25).

calculus, m., S, *calculation* (90, 19, n).

capitulum, neut., *chapter* (74, 21, n).

caput, n., S, *chapter* (74, 21).

casus, m., S, *event, instance* (74, 14).

causa, f., S, *the reason, the explanation* (130, 4).

census, m., S, *the second power of the unknown* (128, 2) ; *quantity* (146, 34).

centenarius, adj., *one hundred* (66, 17).

centenus, adj., *one hundred* (66, 17, n).

certus, adj., *definite* (120, 27).

circumdo, v., *encompass, surround* (78, 12, n).

circumduccio (-tio), f., *perimeter* (78, 13).

circumiaceo, v., S, *surround* (134, 18).

coequacio (coaequatio), f., *equality* (70, 2).

coequalis (coae-), adj., *equal* (88, 1, n).

coequo (coae-), v., *equal* (68, 7).

cognitus, adj., *known* (122, 25).

collectio, f., *sum* (70, 30).

colligo, v., *combine, sum up* (68, 6).

combinatio, f., S, *combination* (128, 2).

committo, v., R, *compare, divide* (126, 2).

compendiose, adv., *briefly* (88, 23).

complector, v., S, *include, embrace* (90, 24).

compleo, v., *fill up, complete* (72, 28) ; *complete* by transferring negative term (108, 4).

completus, adj., *complete* (108, 20).

compono, v., *compose, make* (66, 12).

compositus, adj., S, *composite*, used of numbers, including both tens and units, such as 16 or 24 (136, 13).

comprehendo, v., *include* (112, 17).

comprobo, v., *prove, confirm* (80, 11, n).

concipio, v., *think (of), conceive* (74, 13).

concretus, adj., *formed, complete* (100, 16, n).

conduco, v., *employ, hire* (124, 4).

coniungo, v., *connect, join, add* (68, 2).

considero, v., *consider, reflect* (66, 10).

consimilis, adj., *equal* (80, 6, n).

constituo, v., *constitute, make* (74, 18).

contineo, v., *contain* (84, 16).

contingo, v., *reach, attain* (96, 4, n).

contra, adv., S, *opposite* (78, 25, n).

conuentio, f., S, *custom* (156, 27).

conuersio, f., *reduction* of an equation to simpler form (72, 8).

conuerto, v., *arrive* (66, 21, n) ; *reduce* (68, 23).

costa, f., S, *side* (144, 21).

cum, prep., S, *by* (68, 3) ; *with* (68, 23).

decenarius, adj., *ten* (66, 14, n).

decenus, adj., *ten* (66, 14, n).

declaratio, f., S, *exposition* (128, 1).

deduco, v., with in, *multiply* (86, 15, n).

deficio, v., S, *fail, be lacking* (132, 4).

deleo, v., S, *efface, cancel* (80, 10).

demo, v., with a, *subtract* (110, 9, n).

demonstratio, f., S, *demonstration* (66, 2).

demonstro, v., *designate, represent* (68, 17).

denarius, adj., *ten* (100, 27).

denarius, m., S, *unit of money, penny* (124, 4) ; *unit*, S (128, 6).

descripcio (-tio), f., *description, explanation* (88, 22, n).

designo, v., *represent, designate* (72, 14).

deuinco, v., S, *surpass, exceed* (84, 1).

differentia, f., S, *difference* (150, 19).

differo, v., *differ* (112, 27).

difficultas, f., *difficulty* (88, 24).

digitus, m., S, *digit, unit* (136, 15).

dimensio, f., S, *measurement, side* (84, 4).

dimidium, neut., S, *half* (76, 12).

diminucio (-tio), f., *diminution, subtraction* (74, 23).

diminuo, v., with ex, *subtract* (72, 22), see (72, 2, n) ; with ab, *subtract* (72, 2, n).

diminutiuus, adj., *negative, subtracted* (90, 13, n).

diminutus, part. and adj., *reduced* (68, 4) ; *lessened, negative* (96, 23, n).

dinosco, v., *distinguish, represent* (72, 4).

dinotus, adj., *pointed out, distinguished* (120, 24).

disciplina, *discipline, art, study* (88, 25).

dispono, v., *arrange* (66, 13).

distinctus, adj., *distinct, separate* (70, 22).

distinguo, v., *distinguish, separate, discriminate* (70, 20).

diuido, v., with in, S (100, 5), with per (100 5, n), and with super (100, 8, n), *divide by*; diuido in duo media, *bisect* or *take half of* (86, 13); diuido in duo, *separate into two parts* (102, 21); with inter, *distribute among* (118, 18); diuido per medium, *take the half of* (70, 31).

diuisio, f., *division* (84, 4, n); S, *distribution* (156, 18).

diuisor, m., S, *divisor* (152, 20).

do, v., *give* (74, 11).

doceo, v., *teach, show* (88, 22, n).

dragma (drachma, S), f., *unit* (70, 28); *unit* of money (124, 4, n).

dubito, v., *consider, doubt, be in doubt* (122, 6).

duco, v., with in, *multiply* (80, 8, n); S, *draw* (82, 21).

duplatio, f., S, *doubling* (98, 15).

duplex, adj., *twofold, double* (118, 2).

duplicacio (-tio), f., *doubling, multiplication* (66, 16).

duplico, v., *double* (66, 15); *multiply* (90, 7, n).

duplo, v., S, *double* (140, 33).

duplum, neut., S, *double* (142, 19).

e contrario, S, *on the contrary, in the reverse way* (100, 8).

e conuerso, *in the reverse way* (100, 8, n).

edo, v., *give out, publish* (66, 9).

elicio, v., *draw forth, solve* (108, 29, n); *take* square root, S (148, 4).

elucesco, v., *shine forth* (102, 18).

equalis (ae-), adj., *equal* (78, 4, n).

equidistans (ae-), adj., *equidistant* (82, 8, n).

equiparo (ae-), v., *equal* (68, 20).

equo (ae-), v., *equal* (70, 18).

ergo, adv., *therefore, then, consequently* (68, 6, n).

erigo, v., S, *erect* a perpendicular, *raise up* (132, 15).

error, m., *error* (102, 12).

essentialiter, adv., *essentially, naturally* (66, 13, n).

et, conj., *and* (66, 7); *plus* (90, 16).

euacuo, v., *cancel, vacate* (80, 10).

euenio, v., *come out, result* (82, 25).

exaequo, v., S, *equal* (106, 4).

examen, neut., S, *testing* (116, 7, n).

excedo, v., *exceed, go beyond* (66, 14).

excresco, v., *grow, increase* (66, 19).

exemplar, neut., *model, pattern* (102, 12).

exemplum, neut., *example, type* (122, 7).

exeo, v., *come out* (104, 30).

exerceo, v., *exercise, practise* (102, 18).

exercitium, neut., S, *exercise, problem* (144, 29).

exhibeo, v., *represent, show* (74, 13).

exigo, v., *require, demand* (98, 9).

expedio, v., *carry through* (120, 19).

explano, v., *explain* (86, 1, n).

expono, v., *explain, set forth* (88, 22).

expositio, f., *explanation, exposition* (104, 16).

exprimo, v., *represent, show the form of* (72, 23).

extendo, v., *extend, reach* (72, 12).

extraho, v., with ex, *subtract* (82, 27).

extremitas, f., *end* of a line (78, 23).

extremum, neut., S, *extremity, end* (130, 18).

facile, adv., *easily* (120, 19).

facilis, adv., *easy* (88, 25).

facio, v., *make* (66, 16).

figura, f., *figure* (88, 23, n).

finio, v., *terminate, come to* (104, 27).

forma, f., S, *form* (136, 11).

formula, f., *rule, formula* (80, 11).

fractio, f., S, *fraction* (92, 7).

fractus, adj., S, *fractional* (100, 3).

genero, v., *generate, produce* (90, 23).

genus, neut., *class, species, kind* (70, 22).

geometrice, adv., *geometrically* (76, 20, n).

geometricus, adj., S, *geometrical* (66, 7).

gnomon, m., S, *gnomon*, or the form of a carpenter's square, consisting of three rectangles lying around any given rectangle and forming with it a larger, similar rectangle (132, 21).

habeo, v., *have, constitute* (70, 15).

hypothesis, f., S, *hypothesis* (146, 19).

igitur, conj., *therefore* (68, 2).

ignoro, v., *not to know, be ignorant of* (76, 23).

ignotus, adj., *unknown* (78, 11).

imperfectio, f., *incompleteness* (78, 13).

imperfectus, adj., *incomplete, imperfect* (80, 5).

in, prep., *in* (70, 31) ; with verbs of division or multiplication, *by* (68, 3, n).

inaequalis, adj., S, *unequal* (78, 10).

incertus, adj., *doubtful, unknown* (120, 28, n).

incognitus, adj., *unknown* (122, 14, n).

incurro, v., *incur, run into* (122, 7).

indigeo, v., *need, require* (66, 11).

infinitus, adj., *infinite* (66, 21).

infra, adv. and prep., *below, less than* (68, 4).

inquisicio (-tio), f., *problem* (122, 1).

inscribo, v., *inscribe* (82, 12).

insuper, adv., *in addition, besides* (118, 19).

integer, adj., *integral, whole* (98, 10).

integro, v., S, *make whole* (150, 21).

intelligo, v., *understand, comprehend* (88, 24).

intentum, neut., S, *result proposed* (128, 16).

inter, prep., *between* (84, 17).

interrogacio (-tio), f., S, *question* (70, 31).

inuenio, v., *discover, find* (66, 10).

inuestigacio (-tio), f., *investigation, finding out* (66, 21).

inuestigo, v., *track out, investigate, seek after, look into* (74, 11).

inuicem, adv., *in turn, alternately* (68, 7).

iungo, v., *join, add* (112, 3).

iuxta, prep., *according to, after the manner of, in case of* (68, 29).

lanx, f., *scale* of a balance (96, 23, n).

latitudo, f., *breadth* (78, 5).

latus, neut., *side* (76, 23).

liber, m., *book* (66, 8).

linea, f., *line* (82, 23).

liqueo, v., *appear, be evident* (76, 17).

locus, m., S, *place* (76, 24).

longitudo, f., *length* (78, 5).

magnus, comp. maior, maius, adj., *great* (76, 3, n and 78, 13).

magul, Arabic, *unknown* (122, 12, n).

maneo, v., S, *remain, abide* (72, 2).

manifestus, adj., *evident* (72, 14).

medietacio (-tio), f., *halving, half* (116, 19).

medietas, f., *half* (70, 8).

medio, v., *halve* (74, 26).

medium, neut., *middle, half* (70, 31).

medius, adj., *middle, mean, half* (70, 18).

mensuro, v., *measure* (68, 27).

millenarius, adj., *thousand* (66, 19).

millenus, adj., *thousand* (66, 19, n).

minor, minus, comp. of parvus, adj., *less* (68, 22).

minucia (-tia), f., *a small particle; a small part* (118, 21).

minuendus (minuo), *to be subtracted, negative* (90, 15).

minuo, v., with ex, *subtract from* (74, 14, n).

minutus, adj., S, *negative* (138, 17).

modus, m., *manner, fashion* (66, 15).

multiplicacio (-tio), f., *multiplication* (90, 10).

multiplico, v., *multiply* (66, 17) ; with in, *multiply by* (68, 3, n) ; with cum, S, *multiply by* (68, 3).

multitudo, f., S, *multitude, a great number* (68, 29).

nascor, v., *arise, spring forth* (82, 16).

natura, f., *nature* (98, 9).

naturaliter, adv., *naturally* (96, 29).

necessario, adv., *necessarily* (86, 8).

necesse, adj., *necessary* (78, 25).

negligo, v., S, *neglect, nullify* (96, 21).

nihil, nil, neut., *nothing* (66, 12).

nodus, m., *node, a multiple of ten* (90, 8).

nomino, v., *name, call* (122, 20).

nosco, v., *know, recognize* (72, 4, n).

noticia (-tia), f., S, *axiom* (132, 19).

notus, adj., *known* (120, 27) ; *rational*, S (140, 33).

nullus, adj., S, *not any, null, void* (74, 29).

numerus, m., *number* (66, 9) ; numerus diminutus, *fraction* (98, 10).

nuncio, v., *announce, report* (122, 13, n).

nuncupo, v., *name, call* (120, 23, n).

obtineo, v., *maintain, prove, have, obtain* (78, 6, n).

omnino, adv., *wholly* (102, 17).

omnis, adj., *all, every* (66, 11).

operor, v., *work, operate* (100, 20).

opifex, m., *worker, student* (108, 28).

oppono, v., *set in opposition to, oppose* (96, 4, n).

oppositio, f., *opposition, balancing* (66, 8).

oppositus, adj., *opposed* (122, 16).

ordo, m., *arrangement, denomination* (66, 20).

orior, v., *arise, appear, spring up* (70, 21).

ostendo, v., *represent, show, reveal* (68, 21).

parallelogrammum, neut., S, *rectangle* (82, 8).

pario, v., *produce, obtain* (78, 16, n).

pars, f., *part* (68, 16).

particula, f., *small part, fractional part* (100, 7).

pateo, v., S, *be clear, follow* (128, 26).

paucior, comp. of paucus, adj., *fewer, less* (72, 6, n).

paucitas, f., *fewness, scarcity, paucity* (68, 29).

perduco, v., *bring* to, *lead* to (104, 10).

perfectio, f., *perfection, completion* (80, 22).

perfectus, part. and adj., *complete* (72, 28).

perficio, v., *complete, make* (78, 17).

permaneo, v., *remain* (96, 4, n).

perpendicularis, adj., *perpendicular* (82, 21).

perspectus, adj., *evident, clear* (88, 24).

pertineo, v., *concern, relate to* (76, 19).

peruenio, v., *arrive at, come up to* (66, 17).

pluralitas, f., *many, plurality* (68, 29, n).

plures, adj., the plural of plus, *more* (70, 2).

plus, S, comp. of multus, adj., *plus, more than* (76, 3).

pondus, neut., *weight* (124, 16).

pono, v., *place, assume, assert, propose* (82, 9) ; with super, *place upon* (84, 1).

portio, f., S, *portion, segment* (82, 22).

possum, v., *be able* (102, 12).

postea, adv., *then, afterwards* (66, 22).

posterior, comp. of posterus, adj., S, *latter, following* (76, 16).

postremus, super. of posterus, adj., *last* (118, 23).

praeter, adv. and prep., S, *minus* (136, 17) ; *negative* (138, 7).

precium (pretium), neut., *price* (124, 5).

prefatus (prae-), adj., *before-mentioned* (122, 10).

premitto (prae-), v., *place before* (70, 21).

pretaxo (prae-), v., *mention, assign, enumerate* (74, 14, n).

primum, primo, adv., *first* (74, 13).

primus, adj., *first* (76, 21).

principium, n., *beginning, commencement* (76, 15).

prius, adv., *before, previously* (72, 28).

probatio, f., *trial, test, proof* (76, 23).

probo, v., *try, check, test, prove* (76, 21, n).

procreo, v., *create, produce* (74, 12, n).

produco, v., S, *produce, give* (70, 32).

productum, neut., S, *product* (80, 3).

profero, v., *bring forward, produce, give* (122, 12).

progenero, v., *generate, produce* (106, 12, n).

proicio, v., *cast out, take away* (96, 4, n).

pronuncio, v., *mention, relate* (70, 31).

propono, v., *propose, set forth* (68, 26).

proporcio (-tio), f., *connection, proportion, ratio* (68, 2).

proposicio (-tio), f., *proposition* (76, 21).

propositus, part., *proposed* (102, 14, n).

protendo, v., *extend* (70, 30).

protraho, v., S, *draw, extend* (132, 15).

prouenio, v., S, *come forth, appear* (100, 16).

prout, adv., *just as, as* (76, 17).

punctum, neut., *point* (82, 20).

quadratum, neut., S, *square* (76, 23).

quadratus, adj., *square* (82, 14, n).

quadrilaterus, adj., S, *four-sided, quadrilateral* (82, 14).

quantitas, f., *amount, quantity* (78, 11).

quantum, adv., *how much* (76, 19).

quantus, adj., *how great, as great* (76, 8, n).

quaternarius (sometimes quarternarius, S), adj., *four* (68, 20).

quero (quae-), v., *inquire, ask* (70, 3).

questio (quae-), f., *question, problem* (72, 15).

quociens (quotiens), adv., *how many times* (90, 7).

quotquot, adj., *however many* (72, 6).

radix, f., *root* (72, 1) ; *unknown* (68, 1).

rectangulum, neut., S, *rectangle* (82, 10).

rectangulus, adj., S, *rectangular* (82, 8).

rectus, adj., S, *straight* (as noun 82, 23) ; *right* angled (82, 15).

reddo, v., *give back, make, render* (80, 5).

redeo, v., S, *return, arise* (152, 20).

reduco, v., S, *bring back* (146, 8) ; *multiply* (128, 37).

refero, v., S, *represent* (82, 10).

regula, f., *rule* (72, 18).

relinquo, v., S, *leave* (84, 5).

remaneo, v., *remain, be left* (72, 2, n).

reperio, v., *find, discover* (66, 13).

repeto, v., *repeat* (90, 10).

res, f., *thing* (68, 3) ; *unknown* first power (82, 5).

reseco, v., *cut off* (86, 9).

residuus, neut., *remaining* (76, 16).

respicio, v., *look back, refer* (122, 15).

restauracio (-tio), f., *restoration, transference* of negative terms to the other side of the equation (66, 8).

restauro, v., *restore, transfer* (104, 3).

resto, v., *remain* (80, 16, n).
rumbus (for rhombus), m., *square* (76, 23, n).

scientia, f., *knowledge, science* (66, 10).
scio, v., *know, understand* (74, 26).
seco, v., S., *cut* (132, 16).
secundum, prep., *according to, following* (66, 20).
secundus, adj., *second, following* (120, 25).
semel, adv., *once, a single time* (82, 15).
semis, m., S, *one-half* (142, 3).
semper, adv., *always* (68, 6).
significo, v., S, *signify, represent* (78, 17).
signo, v., *mark, represent, signify* (78, 17, n).
similis, adj., *similar, equal* (82, 9, n).
similiter, adv., *in like manner, similarly* (68, 23).
similitudo, f., *similitude, likeness* (70, 29); ad similitudinem, *likewise* (68, 18).
simul, adv., *at the same time, together, also* (74, 24).
sine, prep., *minus, negative* (90, 22).
singularis, adj., S, *one by one, each* (78, 7).
solucio (-tio), f., *solution* (74, 13).
solus, adj., *alone, pure* (68, 1).
structura, f., S, *construction* (132, 17).
studium, neut., *study, zeal* (102, 18).
sub, prep., *under* (102, 14).
subabiectio, f., S, *subtraction* (134, 33).
subiectus, adj., *accompanying, adjacent* (88, 22, n).
substantia, f., *second power of the unknown* (68, 1); *quantity* (106, 9, n).
subtractio, f., S, *subtraction* (84, 5).
subtraho, v., with ex, *subtract from* (78, 23); with ab, *subtract from* (90, 9).
sufficienter, adv., *sufficiently* (76, 19).
summa, f., *amount, sum* (72, 20, n).
sumo, v., *assume, take* (100, 24).
super, adv. and prep., *above* (72, 1, n); in additions, *to*.
superaddo, v., *add* (104, 15).
superficies, f., S, *area* (136, 3).
superfluum, neut., S, *excess* (150, 19).

superius, comp. adv., *above* (76, 17).
supersum, v., *remain* (92, 19).
supplementa, neut., S, *supplementary rectangles* cut off from a larger rectangle by two lines parallel to the sides and intersecting on the diagonal (134, 8).
supra, prep. and adv., *above* (68, 4).
surdus, adj., S, *surd, irrational* (140, 33).

tantum, adv., with quantum, *as much as* (92, 13).
tempus, neut., *time* (124, 7).
tendo, v., *extend, represent* (84, 7).
termino, v., *limit, end* (82, 13).
ternarium, neut., *three* (98, 4).
ternarius, adj., (consisting of) *three* (84, 23).
tociens (totiens), adv., *so often, so many times as* (90, 7, n).
tollo, v., *take* square root (98, 8); with ex, *subtract, take away* (110, 13, n).
totus, adj., *whole, all* (66, 11).
tracto, v., *treat, use, handle* (70, 1).
trado, v., *impart, set forth* (120, 19).
transfero, v., *translate* (124, 19).
tribuo, v., S, *add* (96, 20).
triplicacio (-tio), f., *tripling* (66, 16).
tripliciter, adv., *triply* (70, 22).
triplico, v., *triple* (66, 15).
tunc, adv., *then, immediately* (84, 26).

uinco, v., *exceed, surpass* (84, 1, n).
ullus, adj., *any* (68, 2).
ultimus, adj., *extreme, last* (120, 26).
unitas, f., *unity, unit* (66, 12).
unus, num. adj., *one* (68, 19).
unusquisque, pronominal adj., *each one, each* (76, 25), (78, 4).
usque, adv., with ad, *up to, as far as* (66, 14).

venalis, adj., *salable, to be sold* (120, 21).
venio, v., S, *come, arrive* (72, 1).
vero, adv., *truly, certainly* (68, 5).
versor, v., *be situated, lie* (102, 17).
verus, adj., *true* (76, 20).
vis, f., *force, significance* (66, 10).

University of Michigan Studies

HUMANISTIC SERIES

General Editors: FRANCIS W. KELSEY and HENRY A. SANDERS

Size, 22.7 × 15.2 cm. 8°. Bound in cloth

VOL. I. ROMAN HISTORICAL SOURCES AND INSTITUTIONS. Edited by Professor Henry A. Sanders, University of Michigan. Pp. viii + 402. $2.50 net.

CONTENTS

VOL. II. WORD FORMATION IN PROVENÇAL. By Professor Edward L. Adams, University of Michigan. Pp. xvii + 607. $4.00 net.

Vol. III. LATIN PHILOLOGY. Edited by Professor Clarence Linton Meader, University of Michigan. Pp. vii + 290. $2.00 net.

Parts Sold Separately in Paper Covers:

THE MACMILLAN COMPANY

Publishers   64–66 Fifth Avenue   New York

VOL. IV. ROMAN HISTORY AND MYTHOLOGY. Edited by Professor Henry A. Sanders. Pp. viii + 427. $2.50 net.

Parts Sold Separately in Paper Covers:

Part I. STUDIES IN THE LIFE OF HELIOGABALUS. By Dr. Orma Fitch Butler, University of Michigan. Pp. 1–169. $1.25 net.

Part II. THE MYTH OF HERCULES AT ROME. By Professor John G. Winter, University of Michigan. Pp. 171–273. $0.50 net.

Part III. ROMAN LAW STUDIES IN LIVY. By Professor Alvin E. Evans, Washington State College. Pp. 275–354. $0.40 net.

Part IV. REMINISCENCES OF ENNIUS IN SILIUS ITALICUS. By Dr. Loura B. Woodruff. Pp. 355–424. $0.40 net.

VOL. V. SOURCES OF THE SYNOPTIC GOSPELS. By Rev. Dr. Carl S. Patton, First Congregational Church, Columbus, Ohio. Pp. xiii + 263. $1.30 net.

Size, 28 × 18.5 cm. 4to.

VOL. VI. ATHENIAN LEKYTHOI WITH OUTLINE DRAWING IN GLAZE VARNISH ON A WHITE GROUND. By Arthur Fairbanks, Director of the Museum of Fine Arts, Boston. With 15 plates, and 57 illustrations in the text. Pp. viii + 371. Bound in cloth. $4.00 net.

VOL. VII. ATHENIAN LEKYTHOI WITH OUTLINE DRAWING IN MATT COLOR ON A WHITE GROUND, AND AN APPENDIX: ADDITIONAL LEKYTHOI WITH OUTLINE DRAWING IN GLAZE VARNISH ON A WHITE GROUND. By Arthur Fairbanks. With 41 plates. Pp. x + 275. Bound in cloth. $3.50 net.

VOL. VIII. THE OLD TESTAMENT MANUSCRIPTS IN THE FREER COLLECTION. By Professor Henry A. Sanders, University of Michigan.

Part I. THE WASHINGTON MANUSCRIPT OF DEUTERONOMY AND JOSHUA. With 3 folding plates of pages of the Manuscript in facsimile. Pp. vi + 104. Paper covers. $1.00.

Part II. THE WASHINGTON MANUSCRIPT OF THE PSALMS. (*In Press.*)

THE MACMILLAN COMPANY

Publishers   64–66 Fifth Avenue   New York

Vol. IX. The New Testament Manuscripts in the Freer Collection. By Professor Henry A. Sanders, University of Michigan.

> Part I. The Washington Manuscript of the Four Gospels. With 5 plates. Pp. vii + 247. Paper covers. $2.00.

> Part II. The Washington Fragments of the Epistles of Paul. (*In Preparation.*)

Vol. X. The Coptic Manuscripts in the Freer Collection. By Professor William H. Worrell, Hartford Seminary Foundation.

> Part I. A Fragment of a Psalter in the Sahidic Dialect. (*In Press.*)

Vol. XI. Contributions to the History of Science. (*Part I ready.*)

> Part I. Robert of Chester's Latin Translation of the Algebra of Al-Khowarizmi. With an Introduction, Critical Notes, and an English Version. By Professor Louis C. Karpinski, University of Michigan. With 4 plates showing pages of manuscripts in facsimile, and 25 diagrams in the text. Pp. vii + 164. Paper covers. $2.00.

> Part II. The Prodromus of Nicholas Steno's Latin Dissertation on a Solid Body Enclosed by Natural Process within a Solid. Translated into English by Professor John G. Winter, University of Michigan, with a Foreword by Professor William H. Hobbs. With 2 plates of facsimiles, and diagrams.

> Part III. Vesuvius in Antiquity. Passages of Ancient Authors, with a Translation and Elucidations. By Francis W. Kelsey. Illustrated.

Vol. XII. Studies in East Christian and Roman Art.

> Part I. East Christian Paintings in the Freer Collection. By Professor Charles R. Morey, Princeton University. With 13 plates (10 colored) and 34 illustrations in the text. Pp. xii + 87. Bound in cloth. $2.50.

> Part II. A Gold Treasure of the Late Roman Period from Egypt. By Professor Walter Dennison, Swarthmore College. (*In Press.*)

THE MACMILLAN COMPANY

Publishers   **64–66 Fifth Avenue**   **New York**

Vol. XIII. Documents from the Cairo Genizah in the Freer Collection. Text, with Translation and an Introduction by Professor Richard Gottheil, Columbia University. (*In Preparation.*)

SCIENTIFIC SERIES

Size, 28 × 18.5 cm.   4°.   Bound in cloth

Vol. I. The Circulation and Sleep. By Professor John F. Shepard, University of Michigan. Pp. x + 83, with an Atlas of 83 plates, bound separately. Text and Atlas, $2.50 net.

Vol. II. Studies on Divergent Series and Summability. By Professor Walter B. Ford, University of Michigan. (*In Press.*)

University of Michigan Publications

HUMANISTIC PAPERS

Size, 22.7 × 15.2 cm.   8°.   Bound in cloth

Latin and Greek in American Education, with Symposia on the Value of Humanistic Studies. Edited by Francis W. Kelsey. Pp. x + 396. $1.50.

CONTENTS

The Present Position of Latin and Greek, the Value of Latin and Greek as Educational Instruments, the Nature of Culture Studies.

Symposia on the Value of Humanistic, particularly Classical, Studies as a Preparation for the Study of Medicine, Engineering, Law and Theology.

A Symposium on the Value of Humanistic, particularly Classical, Studies as a Training for Men of Affairs.

A Symposium on the Classics and the New Education.

A Symposium on the Doctrine of Formal Discipline in the Light of Contemporary Psychology.

THE MACMILLAN COMPANY

Publishers   64–66 Fifth Avenue   New York

Handbooks of Archaeology and Antiquities

Edited by PERCY GARDNER and FRANCIS W. KELSEY

THE PRINCIPLES OF GREEK ART

By PERCY GARDNER, Litt.D., Lincoln and Merton Professor of Classical Archaeology in the University of Oxford.

Makes clear the artistic and psychological principles underlying Greek art, especially sculpture, which is treated as a characteristic manifestation of the Greek spirit, a development parallel to that of Greek literature and religion. While there are many handbooks of Greek archaeology, this volume holds a unique place.

New Edition. Illustrated. Cloth, $2.50

GREEK ARCHITECTURE

By ALLAN MARQUAND, Ph.D., L.H.D., Professor of Art and Archaeology in Princeton University.

Professor Marquand, in this interesting and scholarly volume, passes from the materials of construction to the architectural forms and decorations of the buildings of Greece, and lastly, to its monuments. Nearly four hundred illustrations assist the reader in a clear understanding of the subject.

Illustrated. Cloth, $2.25

GREEK SCULPTURE

By ERNEST A. GARDNER, M.A., Professor of Archaeology in University College, London.

A comprehensive outline of our present knowledge of Greek sculpture, distinguishing the different schools and periods, and showing the development of each. This volume, fully illustrated, fills an important gap and is widely used as a text-book.

Illustrated. Cloth, $2.50

GREEK CONSTITUTIONAL HISTORY

By A. H. J. GREENIDGE, M.A., Late Lecturer in Hertford College and Brasenose College, Oxford.

Most authors in writing of Greek History emphasize the structure of the constitutions; Mr. Greenidge lays particular stress upon the workings of these constitutions. With this purpose ever in view, he treats of the development of Greek public law, distinguishing the different types of states as they appear.

Cloth, $1.50

GREEK AND ROMAN COINS

By G. F. HILL, M. A., of the Department of Coins and Medals in the British Museum.

All the information needed by the beginner in numismatics, or for ordinary reference, is here presented. The condensation necessary to bring the material within the size of the present volume has in no way interfered with its clearness or readableness.

Illustrated. Cloth, $2.25

GREEK ATHLETIC SPORTS AND FESTIVALS

By E. NORMAN GARDINER, M.A., Sometime Classical Exhibitioner of Christ Church College, Oxford.

With more than two hundred illustrations from contemporary art, and bright descriptive text, this work proves of equal interest to the general reader and to the student of the past. Many of the problems with which it deals — the place of physical training, games, athletics, in daily and national life — are found to be as real at the present time as they were in the far-off days of Greece.

Illustrated. Cloth, $2.50

ON SALE WHEREVER BOOKS ARE SOLD

THE MACMILLAN COMPANY

Publishers   64–66 Fifth Avenue   New York

ATHENS AND ITS MONUMENTS

By CHARLES HEALD WELLER, of the University of Iowa.

This book embodies the results of many years of study and of direct observation during different periods of residence in Athens. It presents in concise and readable form a description of the ancient city in the light of the most recent investigations. Profusely illustrated with Half-tones and Line Engravings.

Illustrated. Cloth, $4.00

THE DESTRUCTION OF ANCIENT ROME

By RODOLFO LANCIANI, D.C.L., Oxford; LL.D., Harvard; Professor of Ancient Topography in the University of Rome.

Rome, the fate of her buildings and masterpieces of art, is the subject of this profusely illustrated volume. Professor Lanciani gives us vivid pictures of the Eternal City at the close of the different periods of history.

Illustrated. Cloth, $1.50

ROMAN FESTIVALS

By W. WARDE FOWLER, M.A., Fellow and Sub-Rector of Lincoln College, Oxford.

This book covers in a concise form almost all phases of the public worship of the Roman state, as well as certain ceremonies which, strictly speaking, lay outside that public worship. It will be found very useful to students of Roman literature and history as well as to students of anthropology and the history of religion.

Cloth, $1.50

ROMAN PUBLIC LIFE

By A. H. J. GREENIDGE, Late Lecturer in Hertford College and Brasenose College, Oxford.

The growth of the Roman constitution and its working during the developed Republic and the Principate is the subject which Mr. Greenidge here set for himself. All important aspects of public life, municipal and provincial, are treated so as to reveal the political genius of the Romans in connection with the chief problems of administration.

Cloth, $2.50

MONUMENTS OF THE EARLY CHURCH

By WALTER LOWRIE, M.A., Late Fellow of the American School of Classical Studies in Rome, Rector of St. Paul's Church, Rome.

Nearly two hundred photographs and drawings of the most representative monumental remains of Christian antiquity, accompanied by detailed expositions, make this volume replete with interest for the general reader and at the same time useful as a hand-book for the student of Christian archaeology in all its branches.

Illustrated. Cloth, $1.50

MONUMENTS OF CHRISTIAN ROME

By ARTHUR L. FROTHINGHAM, Ph.D., Sometime Associate Director of the American School of Classical Studies in Rome, and formerly Professor of Archaeology and Ancient History in Princeton University.

"The plan of the volume is simple and admirable. The first part comprises a historical sketch; the second, a classification of the monuments." — *The Outlook.*

Illustrated. Cloth, $2.25

ON SALE WHEREVER BOOKS ARE SOLD

THE MACMILLAN COMPANY

Publishers   64–66 Fifth Avenue   New York

Printed and bound by CPI Group (UK) Ltd, Croydon, CR0 4YY

16/04/2025

14658539-0004